TRIUMPH

Spitfire Mk3

HANDBOOK

Issued by
**STANDARD-TRIUMPH SALES LTD
COVENTRY, ENGLAND**

A member of the British Leyland Motor Corporation

TRIUMPH SPITFIRE Mk. 3

Introduction

Designed and built *to give long and consistent trouble-free service, your Spitfire Mk. 3 also embodies many safety features, the very presence of which will add to your confidence.*

Read carefully the contents of this book which gives, in the simplest possible terms, information vital to the proper operation, care and regular maintenance of the car.

Should you be unable, or prefer not to carry out the various adjustments and the regular maintenance operations described herein, please make use of the Maintenance Voucher Scheme which is fully described in a separate booklet supplied with the car.

Important

**In all communications relating
to Service or Spares, please quote
the Commission Number
(Chassis Number)
Paint and Trim Numbers**

LOCATION OF COMMISSION AND UNIT NUMBERS

Commission, Paint and Trim Numbers—On L.H. Scuttle Side Panel (May be seen by lifting the bonnet)

Engine Number—On L.H. side of Cylinder Block

Gearbox Number—On R.H. side of Clutch Housing Flange

Rear Axle Number—On Hypoid Housing Flange

Note : L.H. and R.H. refers to Left-hand and Right-hand side of the vehicle from the driving position.

IMPORTANT

IN THE INTERESTS OF SAFETY, THE IMPORTANCE OF MAINTAINING CORRECT TYRE PRESSURES CANNOT BE OVER EMPHASISED. PRESSURES SHOULD BE CHECKED AT LEAST EVERY TWO WEEKS OR 1000 MILES, AND MAINTAINED IN ACCORDANCE WITH RECOMMENDATIONS GIVEN ON PAGE 25

STANPART
Spare Parts Service

Replacement parts are not supplied from the factory direct to the general public, but are directed through Distributors who, in turn, supply their Dealers.

Genuine spare parts are marketed under the trade mark "Stanpart" and carry the same guarantee as the original part. The same high quality material is used and the strictest accuracy maintained during manufacture. You are advised, therefore, to insist on the use of these parts should replacements be necessary. Remember, parts which do not carry the trade mark "Stanpart" will invalidate the guarantee if fitted to your vehicle.

The descriptions and illustrations appearing in this book are not binding. The MANUFACTURER, therefore, reserves the right — whilst retaining the basic features of the Models herein described and illustrated — to make at any time, without necessarily bringing this book up-to-date, any alteration to units, parts or accessories deemed convenient for improvement or for any manufacturing or commercial reason.

List

of

Sections

	Page
Introduction	3
Location of Unit Numbers	4
Controls, Instruments and Indicators	6
Overdrive	12
Safety Harness	13
Seats	14
Locks and Keys	15
Soft Top	17
Hard Top	20
Wheels and Tyres	21
Driving from New	26
Care of Bodywork	28
Cooling System	29
Electrical System	32
Bulb Chart	41
Routine Servicing	46
Running Adjustments	60
Recommended Lubricants	70
Lubrication Chart	72
Lubrication Summary	73
General Specification	74
Index	77

Fig. 2

1. Fuel gauge
2. Lighting switch
3. Tachometer
4. Direction indicator monitor
5. Speedometer
6. Ignition/Starter switch
7. Temperature gauge
8. Lights selector switch
9. Horn push
10. Direction indicator control
11. Windscreen wiper switch
12. Windscreen washer control
13. Gear lever
14. Handbrake lever
15. Heater blower switch

Trip Odometer (22, Fig. 7)

The figures within the aperture above the centre of the dial may be used to record the distance of each journey, providing that the figures are initially set at zero.   This is achieved by turning clockwise the knob (18), Figs. 2 and 3, which extends downwards from behind the instrument.

Odometer (23, Fig. 7)

The figures within the aperture below the centre of the dial show the total mileage of the vehicle and may be used as a guide for periodic lubrication and maintenance.

Main Beam Warning Light (24, Fig. 7)

The indicator at the bottom left-hand side of the dial glows blue when the headlamp main beams are selected and is extinguished when the headlamps are "dipped".

Oil Pressure Warning Light (25, Fig. 7)

The centre indicator at the bottom of the dial glows green when the ignition is switched on and is extinguished when the engine runs in excess of idling speed.   Should the light remain on at normal running speeds, stop the engine and check the level of oil in the engine sump. If this is satisfactory, have the lubrication system checked immediately.

Ignition Warning Light (26, Fig. 7)

The indicator at the bottom right-hand side of the dial glows red when the ignition is switched on and is extinguished when the engine is accelerated.   Should the red light remain on whilst driving, a fault is indicated in the battery charging system which should be rectified without delay.

Air Distribution Controls (27, Fig. 8)

The distribution of hot or cold air is controlled by two flap valves located under the facia and near to the feet of the driver and passenger.   When the flaps are closed, maximum air flow is directed to the windscreen for demisting or de-frosting.   Fully open flaps allow maximum air flow to the area around the feet.

Fig. 7

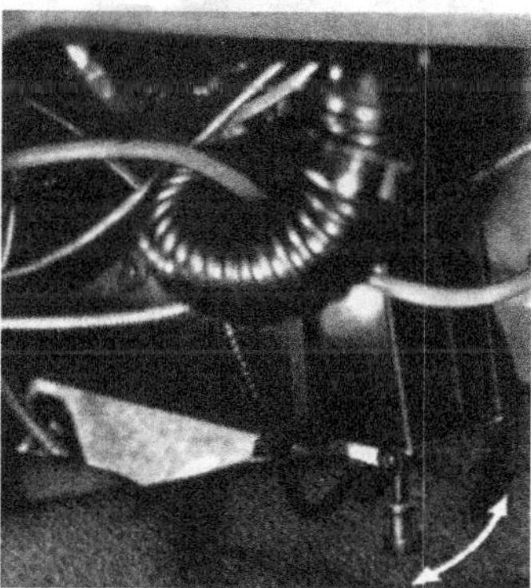

Fig. 8

OVERDRIVE (optional)

An overdrive unit serves as a convenient method of providing, at will, a numerically lower overall gear ratio to reduce engine speed and to improve fuel economy.

The Laycock de Normanville overdrive unit incorporates an epicyclic gear train which is engaged, to give overdrive condition, by a cone clutch moving under the influence of hydraulic pressure generated by a small piston pump. When pressure is released via a control valve, the clutch is returned and held in direct drive by compression springs. A uni-directional roller clutch enables the change into, or out of, overdrive to be made when transmitting full power without loss of road speed.

The hydraulic control valve is linked to an electro-magnetic solenoid, which is operated via a relay by a two-position switch mounted on the steering column (Figs. 9 and 10).

Greatest benefit will accrue from judicious use of the overdrive, the governing factor being that the vehicle continues to run easily without sign of engine labouring, combined with the minimum amount of throttle opening necessary to maintain this condition.

Maximum disengagement speed : 4,800 r.p.m.

The preceding disengagement speed corresponds approximately to peak revolutions in normal gear.

Disengagement of the overdrive at a speed higher than stated may cause damage from "over-revving".

Operation

Move the selector switch down to engage overdrive and up to release it.

Lubrication

The same oil is used for both the overdrive unit and the gearbox, an internal oil transfer hole allowing the flow from the gearbox into the overdrive unit until a common level is attained.

Periodically check and if necessary top up the gearbox-overdrive unit oil level via the gearbox filler orifice. (Refer to page 51).

Fig. 9   Fig. 10

SAFETY HARNESS (optional)

Provision is made for the use of three-point attachment safety belts; anchorage points are built into the vehicle and are shown on Figs. 12 and 13.

Fitting the Harness

Remove the shoulder strap anchor bolt, crimped washer and collar from each wheelarch, pass the bolt through the strap attachment, fit the crimped washer and collar, as shown on Fig. 11, and refit to the vehicle. By means of the latched hooks fit the lap strap to the eye bolts, Fig. 13. The shoulder strap will have a half twist when fitted to the wheelarch, this is the correct position.

Using the Harness

Pass the buckle end of the belt around the hips, and fasten the belt by pushing the locking plate into positive engagement with the buckle. This is denoted by a "click". To release the harness depress the centre panel.

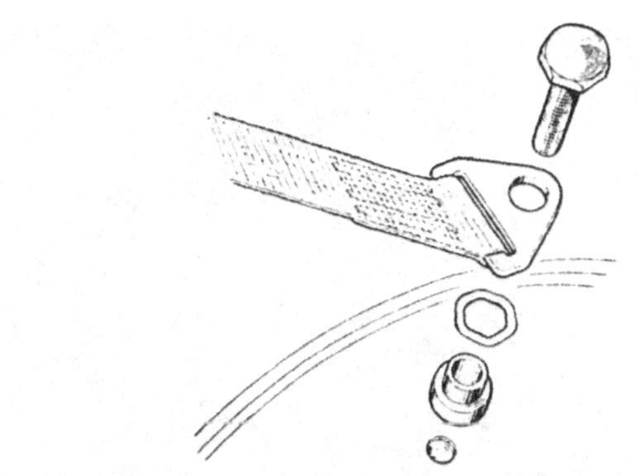

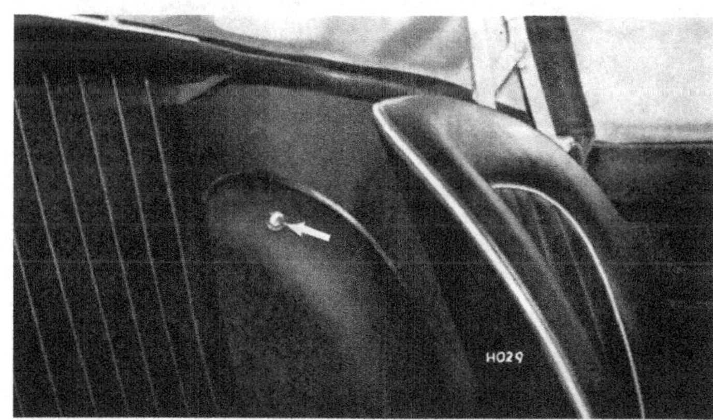

Fig. 11 (*upper*)   Fig. 12 (*lower*)

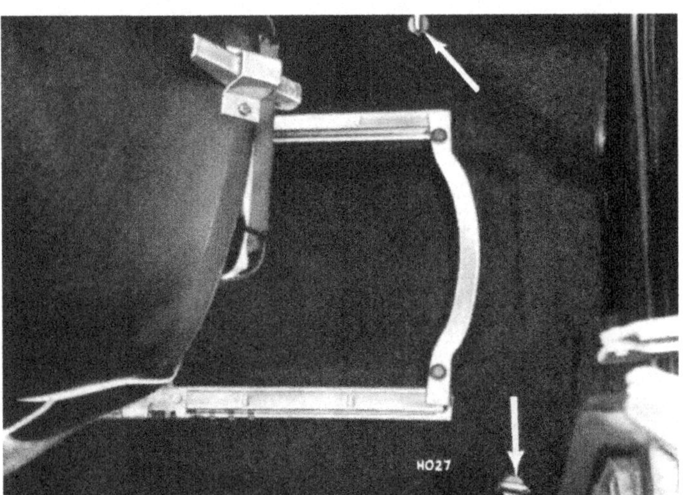

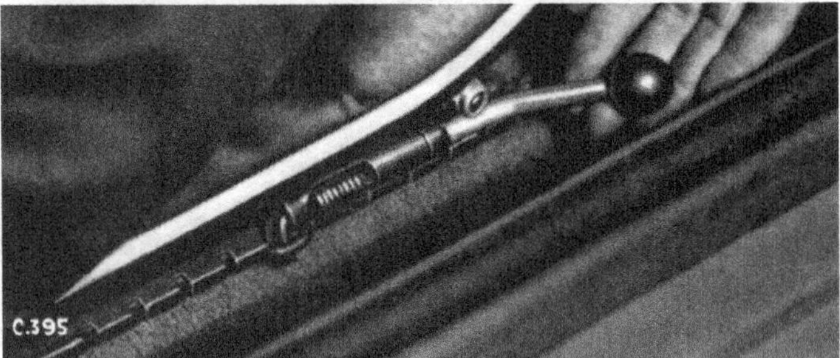

Fig. 13 (*upper*)   **Fig. 14** (*lower*)

Harness Adjustment

The belt should be adjusted so that the hand will pass between the strap and the chest. The lap strap should be reasonably tight and the buckle must rest on the hip nearer the centre of the vehicle.

Adjustment to lower half of belt:

Relieve any tension on the belt and pull the belt over the roller in the buckle, the roller has a self-locking action and once the belt has been adjusted it will lock the belt in position.

Adjustment to upper half of belt:

Pull the grey slide on the lower part of the lap strap upwards to shorten and downwards to lengthen.

Cleaning

Badly stained safety belts may be dry cleaned. The cleaner should be advised of the nature of staining. Belts subjected to normal soiling may be cleaned with soap, or proprietary brand detergents dissolved in hot water.

Seat Adjustment (Fig. 14)

The driver's and passenger's seats are adjustable for leg reach by lifting the lever at the outer side of each seat and sliding the seat to the desired position. Allow the lever to re-engage in the nearest adjustment notch. Both seats will tilt forward to provide access to the rear compartment, when the clip at the base of the seat back is released.

LOCKS AND KEYS

The vehicle is provided with two sets of keys, the spare set being obtained from the selling dealer.   One key is used to operate the ignition, the other is for the luggage compartment and door locks.

Door Locks (Fig. 15)

"Anti-burst" locks are fitted to both doors and are operated by a pushbutton on the outside or by pulling the remote-control lever on the inside.   To lock the door from the inside, push the lever forwards; to lock the door from the outside, insert the key and turn it a quarter turn away from the shut face.   The key will return under the influence of spring loading to a vertical position when it may be withdrawn.

Lubrication

One a month, particularly during freezing weather, apply a few drops of thin machine oil into the latch and key slots.

IMPORTANT.   Do not apply grease to the lock cylinders.

Bonnet Lock (Fig. 16)

The bonnet is opened by raising, as far as possible, a lever on each side to release the catches and lifting the bonnet at its rear.

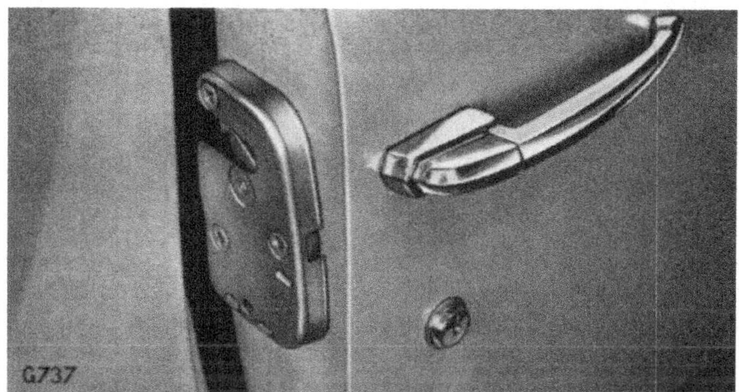

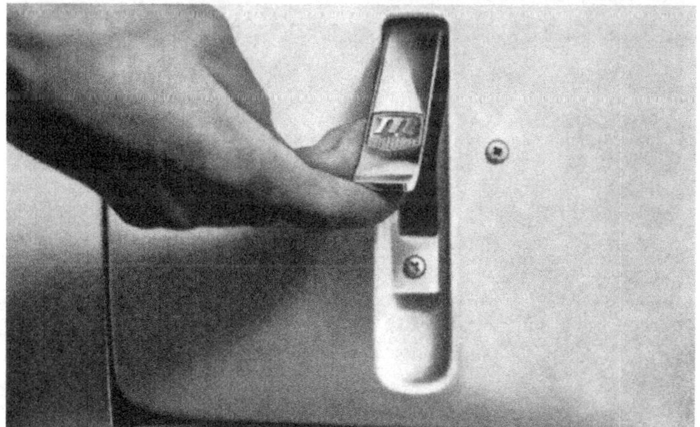

Fig. 15 (*upper*)   **Fig. 16** (*lower*)

This permits the bonnet to pivot on its anchorage to a vertical position where it is held by a folding strut.

To close the bonnet, pull the centre of the strut (arrowed Fig. 17) simultaneously supporting and lowering the bonnet. Press each lever flush with the side of the bonnet to lock.

Luggage Compartment Lid (Fig. 18)

To open the luggage compartment lid, turn the unlocked handle counter-clockwise to a vertical position and raise the lid to its limit before lowering it on to the telescopic support.

Close the lid by raising it slightly to release the catch in the telescopic support, lower, and turn the handle which may be locked by turning the key a half turn counter-clockwise.

Fuel Filler Cap (Fig. 19)

The fuel filler cap, located forward of the luggage locker lid, is opened by depressing a small lever at the side of the cap. Press the cap to close.

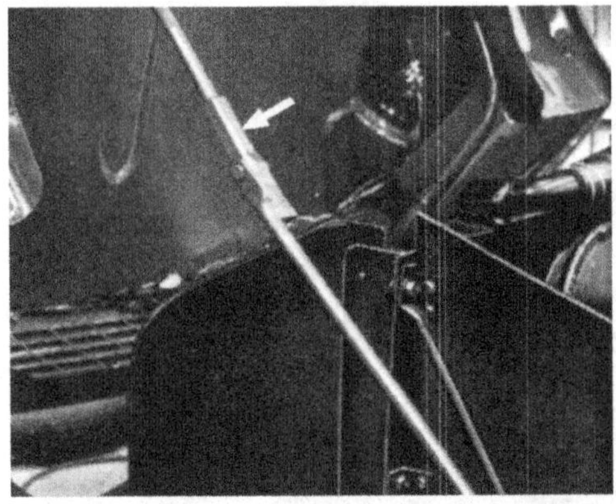

Fig. 17

Fig. 18

Fig. 19

SOFT TOP

The soft top, which is made from P.V.C. material, is supported by a hinged frame. The assembly folds down into the back of the car and is retained in place by a cover.

Raising the Soft Top

Unfasten and remove the hood cover. Fold the sides of the hood fabric outwards and pull the fabric rearwards over the luggage compartment lid. Lifting the front hoodrail, raise the assembly sufficiently to allow the fabric to lie evenly over the frame. Secure the fasteners (four each side, Fig. 22) to the body. Locate the front hoodrail on the windscreen header rail and turn the levers (Fig. 21) inwards. Knock the second hoodstick

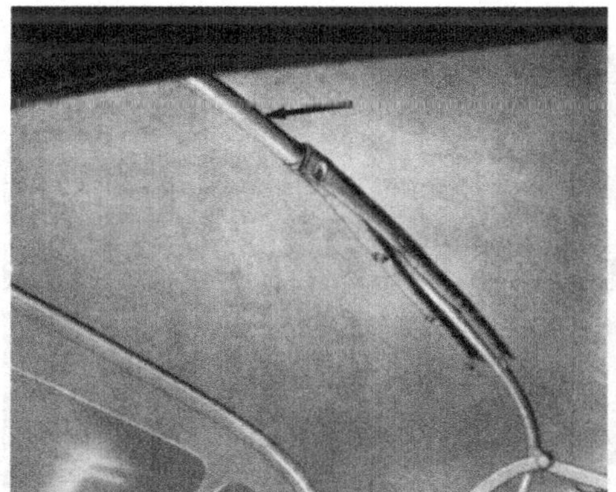

Fig. 20

Fig. 21

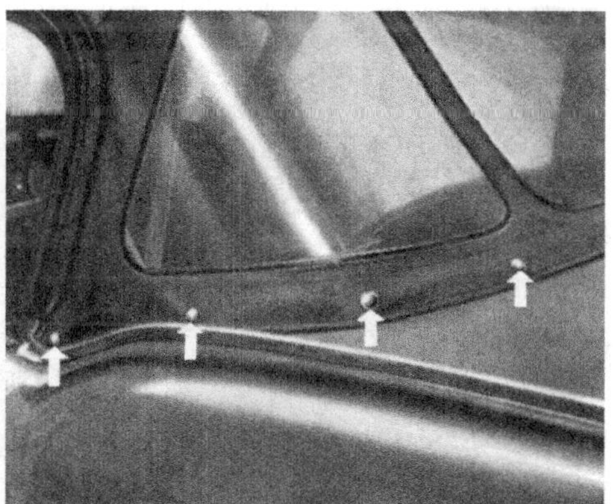

Fig. 22

(arrowed Fig. 20) forwards as far as possible, and secure the fasteners (Fig. 20).

Lowering the Soft Top

Release the fasteners securing the fabric to the second hoodstick (Fig. 20). Release the toggles (Fig. 21) and the fasteners (four each side, Fig. 22) securing the edges of the hood to the body.

Push the front hoodrail rearwards and slightly upwards while knocking the second hoodstick (arrowed Fig. 20) rearwards, until the assembly begins to fold. **DO NOT PULL the second hoodstick downwards.** Continue lowering the frame and pull the fabric flat over the luggage compartment lid (Fig. 23). Fold the fabric forwards over the hoodsticks and turn the sides inwards (Fig. 24). Ensure that the Vybak windows are free from distortion and that the hood fabric is not trapped by the hoodsticks.

Fig. 23

Fig. 24

Retain the hood in position with the cover provided (Fig. 25) as follows:

Attach the cover to the outer fasteners and continue working towards the centre. Attach the inner pillar fasteners and hook the three straps under the bottom hoodstick.

Tonneau Cover (Optional) (Fig. 26)

The tonneau cover provides weather protection for the vehicle interior when the soft top is lowered. It incorporates press-studs for securing to the car and a zip fastener which permits access to either or both of the front seats.

Fig. 25

Fig. 26

HARD TOP (Optional)

The vehicle may be used as an open sports car by removing the hard top assembly as follows:

Unscrew the domed-head bolts securing the hard top side brackets to the door pillar brackets (Fig. 27).

Remove the domed-head bolts and washers securing the hard-top to the windscreen header rail (Fig. 28) and the rear deck panel (Fig. 29).

With the aid of a second operator lift off the hard top assembly

To refit the hard top, reverse the foregoing procedure.

Fig. 27

Fig. 28

Fig. 29

WHEELS AND TYRES

Spare Wheel and Lifting Jack

The tools and spare wheel are housed in the luggage compartment as shown on Figs. 30 and 31.

NOTE.   A variant of the jack illustrated may be provided.

To remove the spare wheel, lift off the cover and unscrew the retaining nut (Fig. 31).

The Jack (Figs. 32 and 33)

Locate the nut of the fixing bolts (rearwards of the front wheel and forwards of the rear wheel) in the head of the jack for safety when lifting a wheel.

Assemble the handle into the jack and turn it to lift the required wheel from the ground.

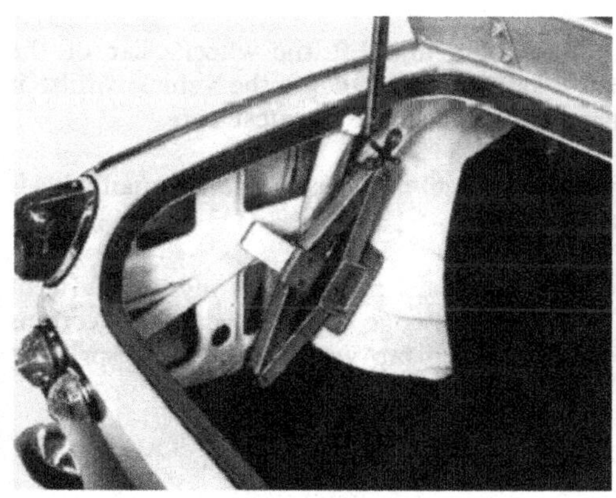

Fig. 30

Fig. 31

Fig. 32

Wheel Changing Procedure (Pressed Steel Wheels)

1. Firmly apply the handbrake and chock the wheel diagonally opposite the one being lifted.

2. Remove the spare wheel from the luggage compartment and make sure that its tyre pressure is correct.

3. Using the special lever provided in the tool kit, lever off the nave plate as shown and slightly loosen the wheel nuts.

4. Place the jack in position and lift the wheel clear of the ground. Should it be necessary to lift the vehicle whilst it is on sloping ground, exercise the greatest care.

5. Completely remove the wheel nuts, exchange the road wheels and replace the nuts.

6. Lower the jack, give the wheel nuts a final tighten and refit the nave plate by placing its edge over the wheel projections and giving the plate a sharp tap with the hand to spring it into position.

Fig. 33 (*left*)   Fig. 34 (*upper*)   Fig. 35 (*right*)

Wire Spoke Wheels (Optional) (Fig. 36)

Before fitting a wheel, check that the adaptor taper (A) and its mating wheel hub taper are undamaged and that each presents a clean painted surface. **DO NOT GREASE THESE SURFACES.** Ensure that the following are undamaged, cleaned and coated with grease: splines (B), screw threads, wheel hub outer taper (C) and the large wheel nut tapers.

Slide the wheel on to the adaptor and while pushing against the wheel hub centre to maintain concentric location, simultaneously screw on the retaining nut by hand until the wheel is felt to seat on the adaptor taper.

Restraining the wheel with one hand, continue tightening by striking the spanner or the ears of the nut with a soft faced hammer. Lower the wheel to the ground and finally tighten. (Figs. 37 and 38).

Check that each wheel retaining nut tightens in the opposite direction to the wheel rotation. The foregoing instructions apply each time a wheel is removed and replaced.

IMPORTANT. Splined adaptors must be fitted to the correct side of the vehicle—left-hand threaded adaptors to the right-hand side and right-hand threaded adaptors to the left-hand side (as viewed from the driver's seat).

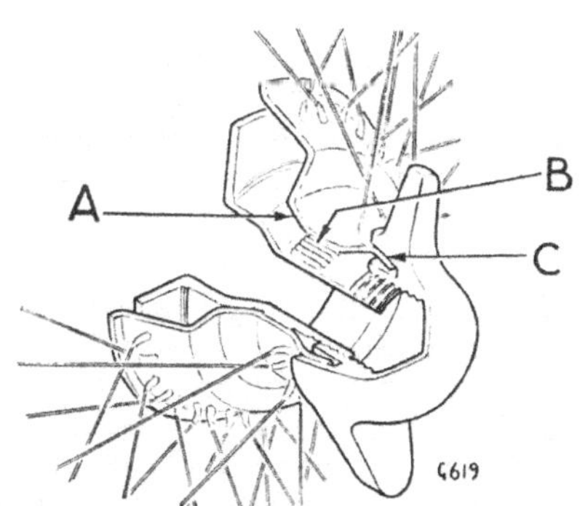

Fig. 36

Fig. 37

Fig. 38

WHEELS AND TYRES

Wheel Alignment

The correct front wheel alignment is $\frac{1}{16}''$ (1·6 mm.) to $\frac{1}{8}''$ (3·2 mm.) toe-in (kerb condition). Excessive misalignment caused by kerb impact or other accidental damage will result in severe tyre wear and faulty steering.

Wheel Run-out and Ovality

The maximum tolerances for both run-out and ovality are as follows:

Press steel wheels	0·070″ (0·18 mm.)
Wire spoke wheels	0·060″ (0·15 mm.)

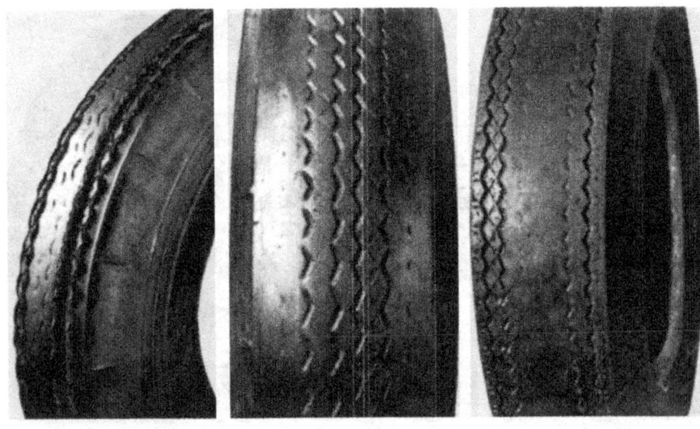

Fig. 39   Fig. 40   Fig. 41

Excessive run-out and ovality will result in severe tyre wear and faulty steering.

Tyre Wear

When new tyres are required it is essential to fit those of the same type. The characteristics of tyres vary considerably and, therefore, all four tyres must be of the same type.

Occasionally remove flints and other road matter from the treads and examine the tyres for sharp fins, flats and other irregularities. An upstanding sharp fin on the edge of each pattern rib is a sure sign of road wheel misalignment (Fig. 39).

Fins on the inside of the pattern ribs indicate toe-in. Fins on the outside edges indicate toe-out. Sharp pattern edges may also be caused by road camber, even when road alignment is correct. In such cases, it is better to make sure by having the track checked with an alignment gauge.

"Spotty" tread wear or flats, can result from grabbing brakes or unbalanced wheel assemblies. Your Standard-Triumph Dealer will check the action of the brakes and re-balance the tyres if required. The original degree of balance is not necessarily maintained, and it may be affected by uneven tread wear, by repairs, by tyre removal and refitting, or by wheel damage and eccentricities. The vehicle may also become more sensitive to unbalance due to normal wear of moving parts.

Excessive wear in the centre of the tread (Fig. 41) results from over-inflation, in which condition the fabric is more easily damaged.

Excessive wear at the outer edges of the tread (Fig. 40) results from under-inflation, a condition which causes excessive heating and premature tyre failure.

Tyres

Adjust tyre pressures in accordance with the recommendations given below.

NOTE. Maintenance of the pressure differential between front and rear tyres is essential for correct steering behaviour.

Never bleed a warm tyre but always adjust the pressure whilst the tyres are cold, i.e. before a run. As the tyres warm up their pressures will increase.

To prolong tyre life, avoid severe braking, sudden changes of direction at speed, and driving over or against high kerbstones, as this can result in severe damage to the tyre walls. Examine the tyres occasionally and remove flints or other road matter which may have become embedded in the treads. Clean off any oil or grease which may be on the tyres by using a cloth moistened in petrol.

Repairs to Tubeless Tyres

A temporary repair can be made to tubeless tyres using a special kit, provided the puncturing hole is small and confined to the central tread area.

The following precautions must be observed:—

1. Do not use more than one plug in each hole.
2. Do not use the tyre for high speeds.
3. Ensure that a permanent "cold patch" or vulcanised repair is made at the earliest opportunity.

RECOMMENDED TYRE PRESSURES

Tyre Size and Type	FRONT		REAR	
	Psi	Kg/cm²	Psi	Kg/cm²
5.20 S — 13	18	1·26	24	1·69
1·45 — 13 Radial ply ..	21	1·47	26	1·83

Where maximum performance is regularly used or if the vehicle is tuned to increase its maximum speed, radial ply tyres are recommended. They are satisfactory at the pressures given above, up to a sustained speed of 110 m.p.h. (176 k.p.h.).

For sustained speeds above 110 m.p.h. (176 k.p.h.) consult the respective tyre company regarding the need for tyres of full racing construction.

DRIVING FROM NEW

Starting the Engine from Cold

Check, and if necessary top up, the radiator water level and the engine oil level. If the car has not been used for several days and fuel has evaporated from the carburettors, refill them by operating the priming lever on the fuel pump. The slight resistance ceases when the float chambers are full.

Apply the handbrake and ensure that the gear lever is in "Neutral". In cold weather pull the choke control fully out; in warm weather pull to the mid-position. In hot climates, do not use the choke. Insert the ignition key and turn it to the "Ignition" position, causing the ignition warning light to glow and the fuel gauge to register the contents of the fuel tank.

From the "Ignition" position, turn the key clockwise against spring pressure to operate the starter motor. Immediately the engine fires, release the key, which will return to the "Ignition" position. Should the engine fail to start at the first attempt, do not re-operate the starter switch until the starter motor has come to rest.

As soon as the engine starts, push the choke control "half in" (cold climates), or fully in (warm climates) and warm the engine at an idling speed of approximately 1,500 r.p.m. This will cause the ignition and oil pressure warning lights to be extinguished, thus indicating satisfactory performance of the generating and lubricating systems. Should either warning light remain on,

stop the engine and establish the cause. Failure to do so may result in serious damage to the engine.

After starting the engine, cylinder wear is minimised if the engine is warmed up quickly by driving away when the oil pressure warning light is extinguished. Do not race the engine to speed up the process but, if possible, maintain an engine speed of approximately 1,500 r.p.m. until the choke can be pushed fully in. In warm climates, use of the choke may be unnecessary. Avoid the use of full throttle during the warming-up period. A thermostat incorporated in the cooling system enables the engine to be warmed up quickly from cold.

Starting with the Engine Warm or Hot

When re-starting a hot engine, depress the accelerator pedal to about one-third of its travel before operating the starter switch. The choke control should not be used.

Running-in

The importance of correct running-in cannot be too strongly emphasized, for during the first 1,000 miles (1,600 km.) of motoring, the working surfaces of a new engine are bedding down.

During this period the valve seats stabilise causing, in some instances, slight distortion and preventing proper seating of a valve. Avoid possible damage resulting from such a condition, by having the compression pressures checked early in the life of the engine after "running-in" is completed. If the pressures are unequal, valve grinding is recommended.

Further attention to the valves should not then be required for a considerable mileage, or until the pressures have again become unequal.

Recommended Speed Limits

Although no specific speeds are recommended during the running-in period, avoid placing heavy loads upon the engine, such as using full throttle at low speeds or when the engine is cold. Running-in should be progressive and no harm will result from the engine being allowed to "rev" fairly fast for short periods provided that it is thoroughly warm and not pulling hard. Always select a lower gear if necessary to relieve the engine of load.

Full power should not be used until at least 1,000 miles (1600 km.) have been covered and even then, it should be used only for short periods at a time. These periods can be extended as the engine becomes more responsive.

Avoid over-revving, particularly in the lower gears. The driver is advised not to drive the car continuously at engine speeds over 6,000 r.p.m., indicated by the beginning of an orange segment on the tachometer, in any gear. However, whilst accelerating through the gears it is permissible to attain 6,500 r.p.m. for short periods, this speed being indicated by the end of the orange segment.

Recommended Fuel

The "Spitfire Mk. III" engine is designed to operate on fuels having a minimum octane rating of 97 (Research Method). Using fuel of this, or higher rating, the static ignition timing should be 6° B.T.D.C.

Where such fuels are not available and if, therefore, it becomes necessary to operate on fuels of lower or unknown rating, the static ignition timing must be retarded just sufficiently to prevent audible detonations (pinking) under all operating conditions. Failure to do so may result in serious damage to the engine.

Driving Hints

Ensure that the gear lever is in neutral before attempting to start the engine.

Do not continue to operate the starter switch if the engine does not fire. This will quickly discharge the battery. Switch off the ignition when the engine is stationary. If left on for long periods, this also will discharge the battery.

CARE OF BODYWORK

Washing

Avoid using a dry cloth to wipe dust from the paintwork and chromium surfaces. Dust is an abrasive and if removed in this way it will scratch the polished surfaces. Wash the vehicle frequently with plenty of running water and a clean soft sponge. Soften and, if possible, remove the mud with water before using the sponge. When all dirt is removed, sponge off and dry with a clean damp chamois leather. Never wash or polish the vehicle under a hot sun.

Removing Grease and Tar

Remove grease or tar with methylated spirits (denatured alcohol). White spirit is also effective, but this must not be applied to rubber, particularly the windscreen wiper blades.

Glass Surfaces

Glass is easily scratched. This can be avoided by always using a damp chamois leather which is especially reserved for use on glass only. If silicone polishes have been used on the body, take care that the polish does not come in contact with the glass. It is extremely difficult to remove and causes the windscreen wipers to smear.

Chromium Plating

Frequent washing and thorough drying is recommended, especially during the winter months when there is likelihood of corrosion through contamination with road salt.

Polishing

After a period of use, the formation of traffic film will cause the paintwork to lose some of its lustre, even though the vehicle has been carefully and regularly washed. The original brilliance may be restored after washing by using a reputable non-abrasive cleaner and polish.

Being the most durable, wax preparations are preferable, but where these are used regularly the old wax must first be removed with a cleaner before further application of new wax. The frequency at which polishing is necessary will depend upon local conditions of air pollution.

Care of Interior, Hoods and Tonneau Covers

Brush and clean the inside of your car each time you wash and polish the outside of it. Use a vacuum cleaner where possible and ensure complete removal of all dust from the interior and trim.

Wash the upholstery (and exterior fabric) with luke-warm non-caustic soapy water. Do not use detergents or household cleaners as these may cause damage. Remove all traces of suds with a clean damp cloth and thoroughly dry the upholstery with a dry duster or towel.

Wipe the facia and instrument panel with a damp cloth only. Wax or other polishes should not be used inside the car.

COOLING SYSTEM

The pressurised "no loss" cooling system incorporates a translucent plastic overflow reservoir (Fig. 45) which collects excess coolant from the radiator as the coolant in the system expands with heat. Depression created as the system cools, causes the coolant to flow back from the reservoir into the radiator. The fluid level which is visible through the translucent reservoir should be maintained at least half full when cold.

Draining

To drain the system, move the heater control (16) Figs. 2 and 3, to the hot position, remove the radiator filler cap Fig. 43, open the tap in the bottom of the radiator (Fig. 42) and the tap at the rear right-hand side of the cylinder block (Fig. 44).

Note. See "Caution", page 47.

Flushing

Efficient cooling is maintained by thoroughly flushing the system once each year before adding anti-freeze. When carrying this out, it is advantageous to remove the drain tap completely and to use plenty of clean running water.

Allowing anti-freeze solution to remain in the system throughout the summer period affords anti-corrosion protection. The solution, however, should be changed at the beginning of each winter period as the inhibitor becomes exhausted.

Fig. 42 (*left*)   **Fig. 43** (*upper*)   **Fig. 44** (*right*)

Filling

Close both drain taps, open the heater control fully and remove the radiator filler cap. Fill the cooling system with clean (soft) water and run the engine at approximately 1,500 r.p.m. for 1 or 2 minutes. Top up the radiator and replace the filler cap. Completely fill the plastic overflow reservoir with clean water.

Screen Washer (Fig. 46)

Examine the water level in the plastic windscreen washer container. If required, lift off the cap and replenish the container with clean water. Under freezing conditions, fill the screen-washer container with a mixture of methylated spirits (alcohol) and water, the recommended proportions being 1 part alcohol to 2 parts water. This may then be used to disperse ice and snow from the windscreen. Do not use anti-freeze solution in the windscreen washer as this may discolour the paintwork and damage the wiper blades and sealing rubber.

Fig. 45

Fig. 46

Frost Precautions

The car heater cannot be completely drained by normal methods. Therefore frost damage will not be prevented by merely draining the radiator.

For your protection during freezing weather, an approved anti-freeze solution should be added to the coolant in the radiator. Because of the searching effect of these solutions, advise your Dealer to check the system for leaks before adding the anti-freeze.

At certain temperatures glycol water solutions adopt a "mushy" state with a viscosity which impairs circulation and can immobilise or damage the water pump. Therefore, consult the following chart before adding anti-freeze, for the degree of frost protection required.

ANTI-FREEZE CONCENTRATION	25%	30%	35%
Complete Protection :— Vehicle may be driven away immediately from cold	10°F (—12°C)	3°F (—16°C)	—4°F (—20°C)
Safe Limit :— Coolant in mushy state. Engine may be started and vehicle driven away after short warm-up period	0°F (—17°C)	—8°F (—22°C)	—18°F (—28°C)
Lower Protection Limit :— Prevents frost damage to cylinder head, block and radiator Engine should NOT be started until thawed out	—14°F (—26°C)	—22°F (—30°C)	—28°F (—33°C)

ELECTRICAL SYSTEM

A 12-volt NEGATIVE earth return system is employed in all circuits shown on Figs. 65 and 66. The system comprises a generator, control unit and battery which supplies current to the ignition system, starter motor, lights and ancillary equipment.

To safeguard against damage resulting from a short circuit, always disconnect the negative cable from the battery before removing or disconnecting an electrical component.

WARNING. If the vehicle is to be equipped with a radio, connected to the car electrical system, ensure that the radio is of NEGATIVE GROUND POLARITY, or serious damage will result. Care must be taken to ensure that replacement equipment is also of the correct polarity.

Battery (Figs. 47 and 48)

The 12 volt lead-acid type battery, housed in the engine compartment, is accessible when the bonnet is raised.

Maintain the top of the battery in a clean condition, and ensure that the battery posts and terminals are clean, tight and protected against corrosion by a coating of petroleum jelly. Maintain the electrolyte at the correct level, by adding distilled water to the cells when necessary. Should acid spillage result

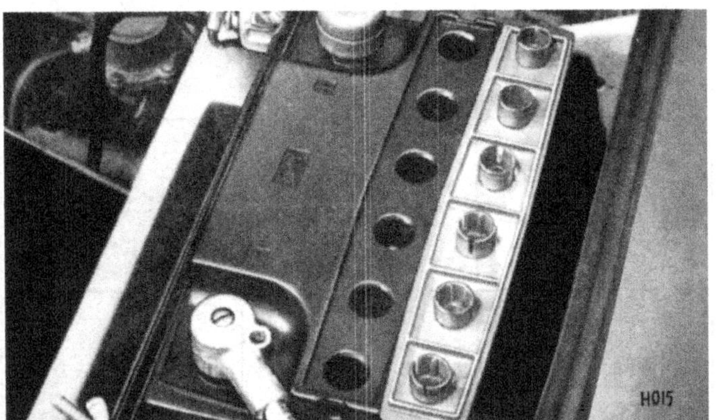

Fig. 47

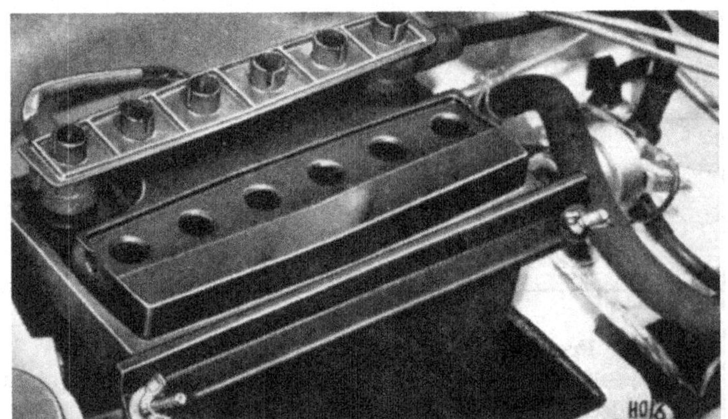

Fig. 48

from this operation, or for any other reason, clean the affected area with an ammonia moistened cloth.

If, for any reason, the battery is completely discharged, or reduced to a low state of charge, it should be recharged as soon as possible; it will rapidly deteriorate if left in a discharged condition.

When renewing or refitting the battery, ensure that it is firmly clamped to its platform.

Generator and Control Box (Figs. 49 and 50)

The generator operates in conjunction with a three bobbin type voltage and current regulator unit which automatically controls the charging rate to suit the needs of the battery. A cut-out device within the regulator unit prevents the battery from being discharged through the generator, when the generator is not charging. In this event, the ignition warning light glows.

The generator, mounted on the front left-hand side of the engine is driven by a Vee-belt which is adjusted for tension by slackening the attachments (1 and 2, Fig. 49), and pivoting the generator to the desired position before retightening the bolts.

The belt should be sufficiently tight to drive the generator and water pump without unduly loading the bearings.

Fig. 49

Fig. 50

Starter Motor

The starter motor, mounted on the rear left-hand side of the engine, is operated by a solenoid switch (Fig. 51) which is remotely controlled by the ignition switch, or manually controlled by pressing the end of the solenoid thus causing the crankshaft to rotate as required whilst working over the engine compartment. The solenoid is attached to the bulkhead adjacent to the battery.

Fig. 51

IMPORTANT.   Before manually operating the starter solenoid make quite sure that the gear lever is in "Neutral" and the handbrake is firmly applied.

Should the starter pinion jam in mesh with the flywheel, it may be released by switching off the ignition, selecting top gear, and rocking the car to and fro, or by turning the squared end of the starter motor shaft in a clockwise direction, viewed from the end of the shaft.

Relays

A relay (arrowed Fig. 51) is incorporated in the horn circuit, and in the overdrive circuit also, when an overdrive is fitted. The relays are mounted adjacent to each other, on the centre of the bulkhead in the engine compartment.   Failure of a relay necessitates renewal of the unit.

Oil Pressure Switch

An oil pressure switch contains a spring-loaded diaphragm which opens a pair of contacts to extinguish a warning light on the facia when the pressure exceeds $4\frac{1}{2}$ to $7\frac{1}{2}$ p.s.i. (0·22 to 0·53 kg/cm^2).   The switch is screwed into the side of the engine and communicates, via a drilling, with the main oil gallery.

Fuses (Fig. 52)

The fuse assembly, located at the left-hand side of the bulk-head, houses three 35 amp. operational fuses and two spares at positions (4).

The top fuse (1) fed by a white cable from the ignition/starter switch protects the reverse lamp circuit, flasher circuit, heater circuit (when fitted), fuel and temperature indication circuits, stop lamp circuit and windscreen wiper circuit.

The middle fuse (2) fed by a red/green cable from the master light switch protects the front parking lamp circuit and tail lamp and plate illumination lamp circuit.

The bottom fuse (3) fed by a brown cable direct from the battery protects the horn circuit and headlamp flasher circuit.

Failure of a particular fuse is indicated when all circuits protected by it become inoperative.   If a new fuse fails, establish the cause and rectify the fault before renewing the fuse.

H.T. Cables

All high tension cables fitted to the ignition system are made from carbon impregnated nylon or cotton cords, encased in rubber or neoprene to form a high resistive conductor. Replacement cables should be obtained from a Standard/Triumph Distributor or Dealer and must be of the same length and type as the originals.

Fig. 52

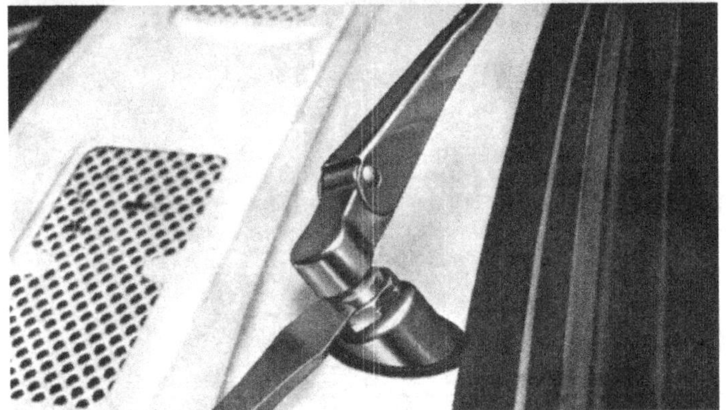

Fig. 53 (*upper*)   **Fig. 54** (*lower*)

Windscreen Wiper (Fig. 53)

A parking switch is incorporated in the cover of the wiper gearbox. On switching off at the wiper control switch, the motor continues to run until the parking switch interrupts the earth return circuit and stops the motor.

The parking position may be adjusted by slackening the screws retaining the cover plate and rotating the domed cover the requisite amount.

Windscreen Wiper Arms

To remove an arm, prise off with a screwdriver (Fig. 54). To refit, push the arm on to the spindle. Check the wiper operation and re-position the arm if necessary.

Windscreen Wiper Blades

To remove a wiper blade, hold the blade firmly and depress the top of the arm (Fig. 55). To refit, push the blade on to the end of the arm.

Temperature and Fuel Gauges

A temperature indicator, comprising a transmitter and gauge unit operates on a 10 volts system controlled by a voltage stabiliser.

The voltage stabiliser is attached by a screw to the rear of the speedometer, and provides a constant 10 volts supply to a fuel gauge and tank unit.

Flasher Unit (Fig. 56)

The flasher unit mounted in a socket attached to the bulkhead below the facia contains an actuating wire which heats and cools alternately to operate a set of contacts. The contacts control the current supply to the selected flasher lamps. A secondary set of contacts controls the flasher warning light which operates when the system is functioning correctly. A defective unit must be renewed. Renewal is effected by pulling the defective unit from the socket and inserting a replacement.

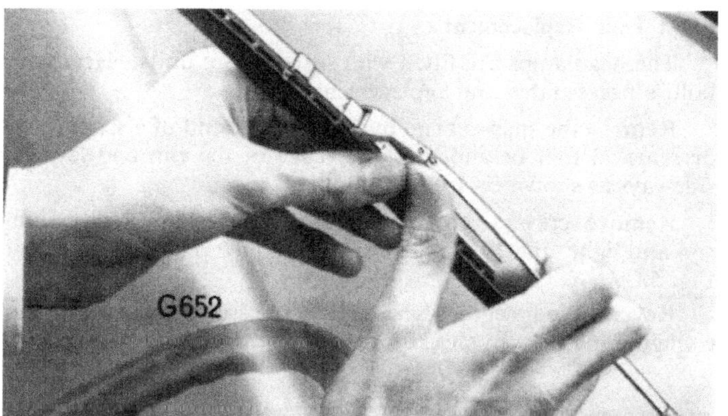

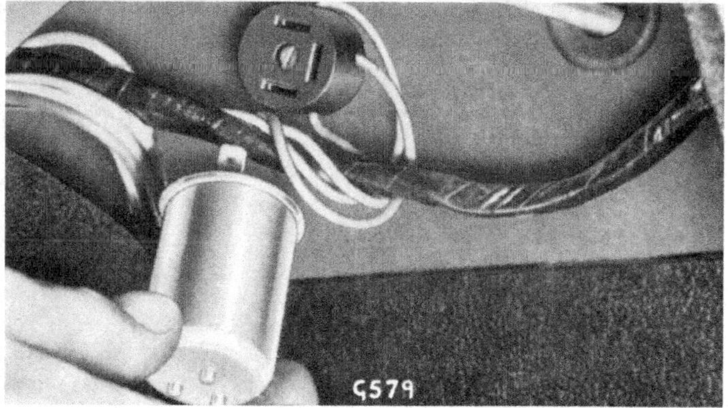

Fig. 55 (*upper*) **Fig. 56** (*lower*)

Light Unit Replacement

The headlamps are fitted with sealed beam units. Headlamp failure necessitates unit replacement.

Remove the snap-on rim by inserting the end of a screwdriver or removal tool behind the lower edge of the rim and levering sideways as shown on Fig. 57.

Remove screws 1, 2 and 3 (Fig. 59) and withdraw the retaining rim and light unit, Fig. 58. Pull the adaptor plug from the unit, Fig. 58.

Reverse the foregoing sequence to re-assemble the lamp, first engaging the bottom clip before pressing the rim home.

Headlamp Alignment (Fig. 59)

The beam is aligned in the vertical plane by turning the screw (B) at the top of the lamp and in the horizontal plane by turning the screw (A) on the side. Alignment of the high beam of one lamp is best carried out with the other lamp covered.

Maximum illumination is obtained, and discomfort to other road users is prevented, by ensuring that the lamp beams do not project above the horizontal when the vehicle is fully laden. When adjustments are necessary, they should be entrusted to a Dealer having suitable beam setting equipment.

Fig. 57

Fig. 58

Fig. 59

Side (Parking) and Direction Indicator Lamps (Front) (Fig. 60)

The side and flasher lamp has two bulbs incorporated in the same housing. The parking bulb is accessible after two screws have been removed from the rim, and the rim and lens lifted away.

Tail Brake/Stop Lamp (Fig. 61)

Take out the screw and lift the lens away to gain access to the bulb.

Rear Direction Indicator (Fig. 62)

Take out both screws and withdraw the lens to gain access to the bulb.

Reversing Lamps (Fig. 63)

With the aid of a thin screwdriver turn back the rubber and remove the rim. This then permits the glass lens to be similarly removed. When re-assembling the components fit the glass lens first.

Fig. 60

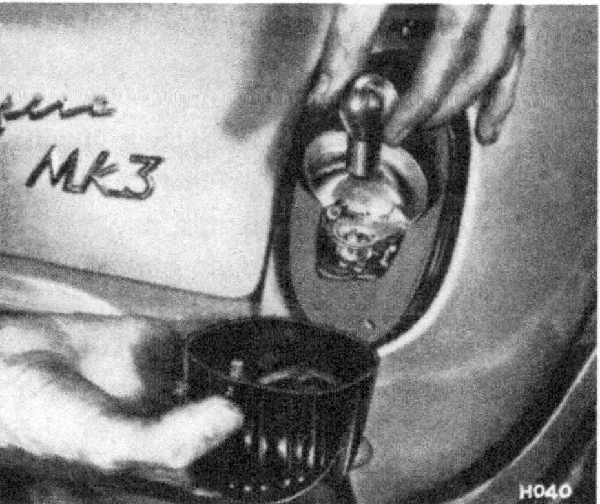

Fig. 61

Fig. 62

Fig. 63 *(upper)*   Fig. 64 *(lower)*

Plate Illumination Lamps (Fig. 64)

Remove the screw securing the cover and lens to the lamp body to gain access to the bulb.

Instrument Illumination and Warning Lamp Bulbs

The bulbs serving instrument illumination and warning lamps, may be renewed by reaching up behind the facia and pulling the bulb-holder from engagement with its housing. The bulbs may then be unscrewed from the holders and replacements fitted.

BULB CHART

Lamp	Watts	Lucas Part No.	Stanpart No.
Headlamps L.H. Dip — Normal Sweden	60/45	54521872	512231*
R.H. Dip — Normal	45/40	410	510218
France Sweden	45/40	411	510219
After 1st September, 1967	45/40	410	510218
U.S.A.	50/40	54522231	—— *
Front parking lamps	6	207	57591
Front flasher lamps	21	382	502379
Tail/Stop lamps	6/21	380	502287
Rear flasher lamps	21	382	502379
Reverse lamps	21	382	502379
Plate illumination lamp — Normal	6	989	59467
U.S.A.	4	222	501436
Instrument illumination	2·2	987	59492
Warning lights	2·2	987	59492

*Sealed beam light unit

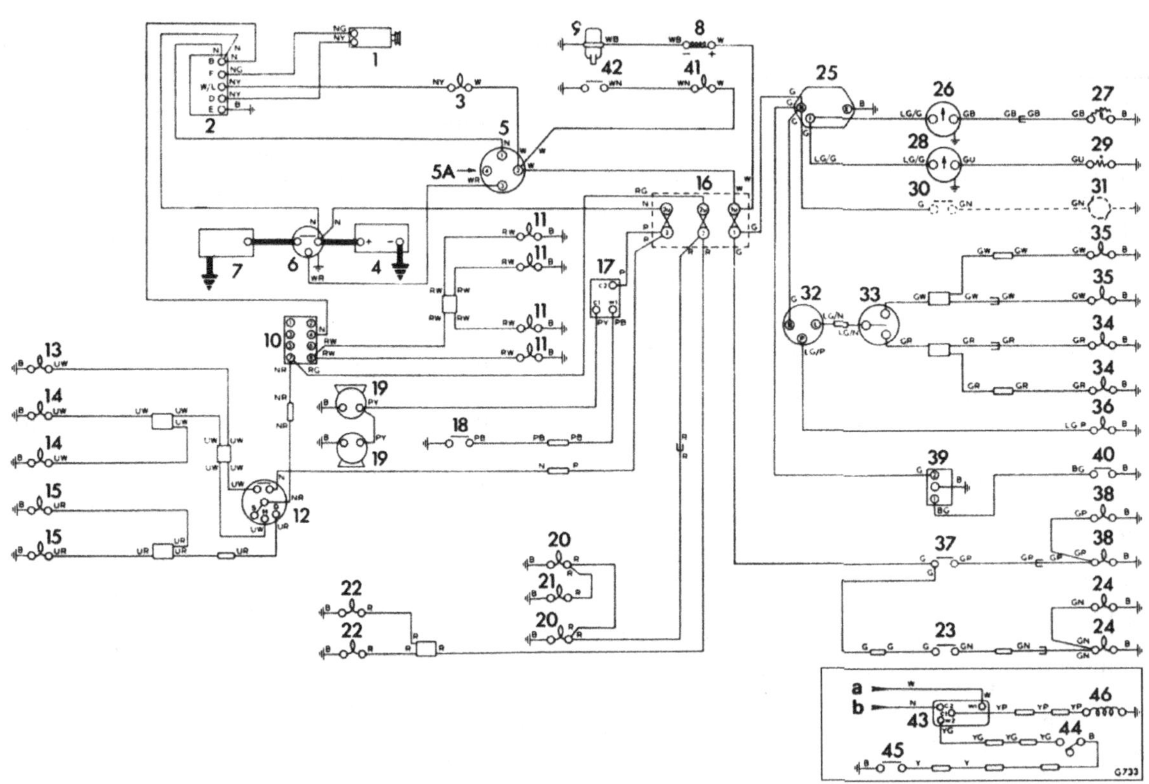

Fig. 65. Wiring Diagram (R.H.S.)

KEY TO FIG. 65

1. Generator
2. Control Box
3. Ignition Warning Light
4. Battery
5. Ignition/Starter Switch
5A. Ignition/Starter Switch—
   Radio Supply Position
6. Starter Solenoid
7. Starter Motor
8. Ignition Coil
9. Ignition Distributor
10. Master Light Switch
11. Instrument Illumination
12. Lights Selector Switch
13. Main Beam Warning Light
14. Main Beam
15. Dip Beam
16. Fuse Assembly
17. Horn Relay
18. Horn Push
19. Horn
20. Tail Lamp

21. Plate Illumination Lamp
22. Front Parking Lamp
23. Reverse Lamp Switch
24. Reverse Lamp
25. Voltage Stabilizer
26. Fuel Indicator
27. Fuel Tank Unit
28. Temperature Indicator
29. Temperature Transmitter
30. Heater Switch (Optional)
31. Heater Motor (Optional)
32. Flasher Unit
33. Direction Indicator Switch
34. L.H. Direction Indicator Lamp
35. R.H. Direction Indicator Lamp
36. Direction Indicator Warning Light
37. Stop Lamp Switch
38. Stop Lamp
39. Windscreen Wiper Motor
40. Windscreen Wiper Switch
41. Oil Pressure Warning Light
42. Oil Pressure Switch

OVERDRIVE (OPTIONAL)

43. Overdrive Relay
44. Overdrive Column Switch
45. Overdrive Gearbox Switch
46. Overdrive Solenoid
a. From Ignition/Starter Switch—
   Connector 2
b. From Ignition/Starter Switch—
   Connector 1

CABLE COLOUR CODE			
N	Brown	LG	Light Green
U	Blue	W	White
R	Red	Y	Yellow
P	Purple	B	Black
G	Green		

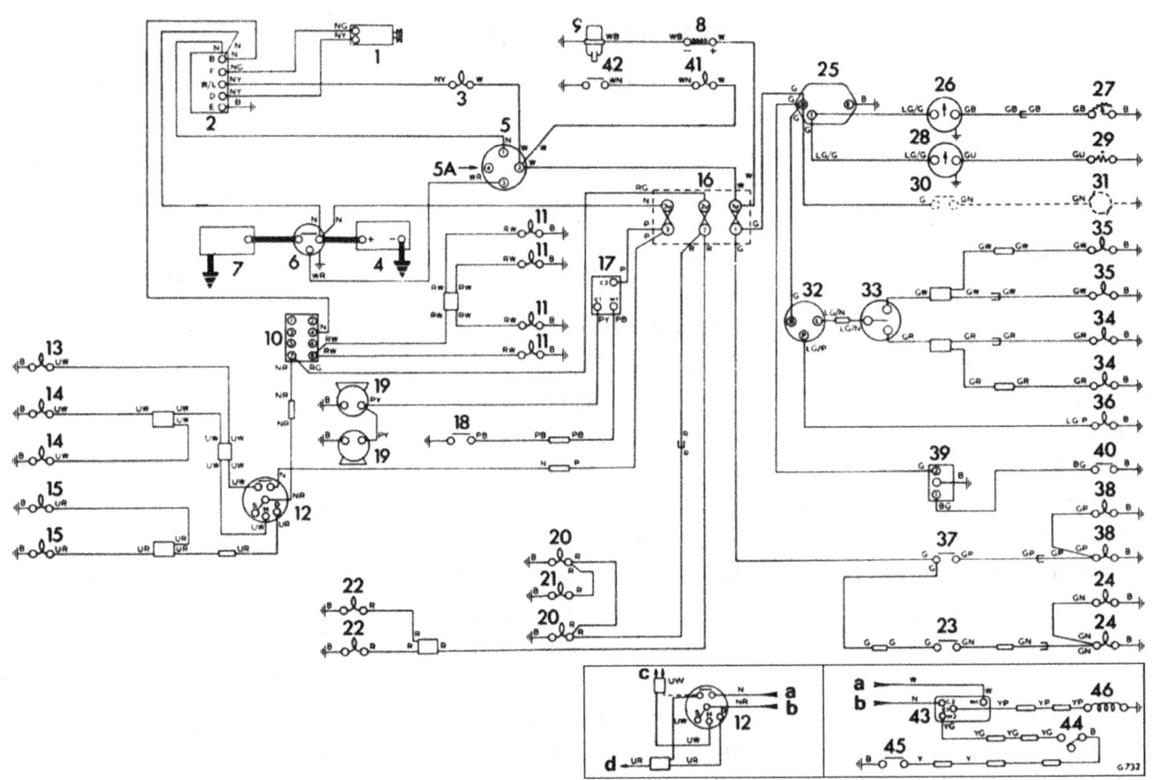

Fig. 66. Wiring Diagram (L.H.S.)

KEY TO FIG. 66

1. Generator
2. Control Box
3. Ignition Warning Light
4. Battery
5. Ignition/Starter Switch
5A. Ignition/Starter Switch—Radio Supply Position
6. Starter Solenoid
7. Starter Motor
8. Ignition Coil
9. Ignition Distributor
10. Master Light Switch
11. Instrument Illumination
12. Lights Selector Switch
13. Main Beam Warning Light
14. Main Beam
15. Dip Beam
16. Fuse Assembly
17. Horn Relay
18. Horn Push
19. Horn
20. Tail Lamp

21. Plate Illumination Lamp
22. Front Parking Lamp
23. Reverse Lamp Switch
24. Reverse Lamp
25. Voltage Stabilizer
26. Fuel Indicator
27. Fuel Tank Unit
28. Temperature Indicator
29. Temperature Transmitter
30. Heater Switch (Optional)
31. Heater Motor (Optional)
32. Flasher Unit
33. Direction Indicator Switch
34. L.H. Direction Indicator Lamp
35. R.H. Direction Indicator Lamp
36. Direction Indicator Warning Light
37. Stop Lamp Switch
38. Stop Lamp
39. Windscreen Wiper Motor
40. Windscreen Wiper Switch
41. Oil Pressure Warning Light
42. Oil Pressure Switch

OVERDRIVE (OPTIONAL)

43. Overdrive Relay
44. Overdrive Column Switch
45. Overdrive Gearbox Switch
46. Overdrive Solenoid
a. From Ignition/Starter Switch—Connector 2
b. From Ignition/Starter Switch—Connector 1

DIP BEAM FLASHER (Italy only)

12. Lights Selector Switch
a. From Fuse Assembly
b. From Master Light Switch
c. To Main Beam Circuit
d. To Dip Beam Circuit

CABLE COLOUR CODE			
N	Brown	LG	Light Green
U	Blue	W	White
R	Red	Y	Yellow
P	Purple	B	Black
G	Green		

ROUTINE SERVICING

The lubricants listed on pages 70 and 71 have maintained a high standard of quality over many years and are approved only after extensive tests in collaboration with the oil companies concerned. In countries where these oils are unobtainable, use similar oils having the same characteristics. The use of only high grade lubricants is vitally important and cannot be over-emphasised.

Engine

When a new car is delivered, the engine sump contains a quantity of special oil, sufficient for the running-in period. Should the level fall below the low mark on the dipstick, the sump may be topped-up with any of the approved lubricants.

At the "Free Service", the running-in oil is drained and the sump replenished to the level of the high mark on the dipstick, with one of the approved oils.

Gearbox, Overdrive and Rear Axle

Rear axles, gearboxes and overdrive units fitted to new cars are filled with a special oil, formulated to give all necessary protection to new gears. **This oil should not be drained** but may be topped up with any of the approved oils.

Braking System

In addition to adjustment and examination/renewal of shoes and pads at the intervals recommended in the following pages, it is strongly recommended by the Royal Society for the Preven-tion of Accidents and the Manufacturers of brake components, that the braking system be overhauled every 36,000 miles (60,000 km.) or 3 years (whichever is sooner).

This involves dismantling the brake system and examining each item for defects. All seals and defective components must be renewed.

Owners are urged to seek the assistance of any Standard-Triumph Distributor or Dealer who will be pleased to estimate for the work, which is of such a nature that it should be entrusted only to skilled workshop personnel.

Preventive Maintenance

To ensure continued efficiency and prolonged vehicle life, the maintenance voucher scheme, produced by Standard-Triumph engineers, offers a carefully designed plan of lubrication require-ments and adjustment checks at pre-determined periods.

Operated by all Standard-Triumph dealers, and specifically recommended to owners wishing to obtain the greatest pleasure from their motoring, the scheme involves the use of a series of Maintenance Vouchers contained in a booklet supplied with the car. Service operations appropriate to mileage or periods of time are listed on pages preceding the vouchers.

The space provided on the counterfoil of each voucher should be filled in by the dealer to constitute proof of regular servicing, should this be required when making a claim under the warranty, or when selling the vehicle.

PERIODIC CHECKS

Engine—Daily

Prior to starting out on a long run, or every 250 miles (400 km.), check the engine oil level and, if necessary, add oil until the level reaches the high mark on the dipstick.

Before checking the level, make sure that the car is standing on level ground. The dipstick, located on the right-hand side of the crankcase, may then be withdrawn, wiped clean and pushed fully home before withdrawing it for reading. Should the level be at the lower mark on the dipstick, 2 pints (imperial) (1·14 litres) (2·4 pints U.S.A.) will be required for topping up via the cap (Fig. 67).

Radiator Water Level—Weekly (Fig. 68)

The level of water, visible through the translucent plastic reservoir mounted forward of the radiator, should be maintained at least "half full" by adding soft water, when required, via the screwed cap.

Should the reservoir be allowed to empty, remove the radiator filler cap, completely fill the radiator, replace the cap and fill the plastic reservoir.

CAUTION. If the engine is hot, avoid danger from scalding by exercising extreme care when removing the radiator filler cap. Turn it a half-turn and allow pressure to be fully released before completely removing the cap.

Fig. 67   Fig. 68

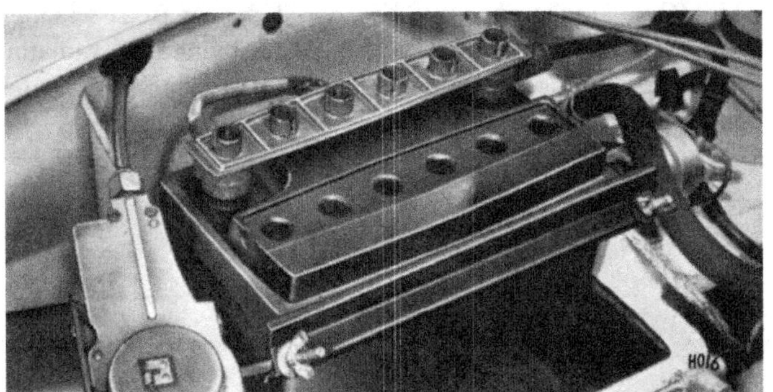

Fig. 69 *(upper)*   Fig. 70 *(lower)*

Screen Washer

Examine the water level in the plastic windscreen washer container.   If required, unscrew the cap and replenish the container with clean water.   (Refer to page 30).

Tyres

The maintenance of correct tyre pressures is an important factor governing tyre life, steering behaviour, braking, and riding comfort.   It is, therefore, important that tyre pressures are checked regularly at periods not exceeding two weeks, and the losses, due to diffusion, are made good.   Correct tyre pressures are given on page 25.

Adjust the pressures whilst the tyres are cold, *i.e.*, before a run. As the tyres warm up their pressures increase.   A warm tyre bled to the recommended pressure will be under-inflated when cold.

MONTHLY CHECKS

Battery (Figs. 69 and 70)

Examine the level of the electrolyte in the cells and, if necessary, add distilled water via the filler orifices to bring the level up to the top of the separators.

The use of a Lucas Battery Filler will aid topping-up. Ensure that the battery filler is filled with distilled water and insert it into a filler plug orifice until it rests gently on top of the separators. Sufficient water will pour into the cell to bring the electrolyte to its correct level.   Check each cell in turn.

CAUTION. Never use a naked light when examining the battery. The mixture of oxygen and hydrogen given off by the battery is dangerously explosive.

Brake and Clutch Master Cylinders (Figs. 71 and 72)
   Wipe the master cylinder caps clean, remove them and check the fluid level in the clutch (2) and brake (1) master cylinder reservoirs.   If necessary, top up the fluid until it is level with the marking on the side of the reservoirs.   Ensure that the breather hole in the raised centre of each cap is unobstructed before refitting the caps.

NOTE.   As the brake pads wear, the level of fluid in the master cylinder falls.   The addition of fluid to compensate for pad wear is unnecessary.   Should the level have fallen appreciably, check the condition of the pads.   If their condition is satisfactory establish the cause of loss and rectify the defect immediately.   Refer to page 68, "Procedure for Bleeding".

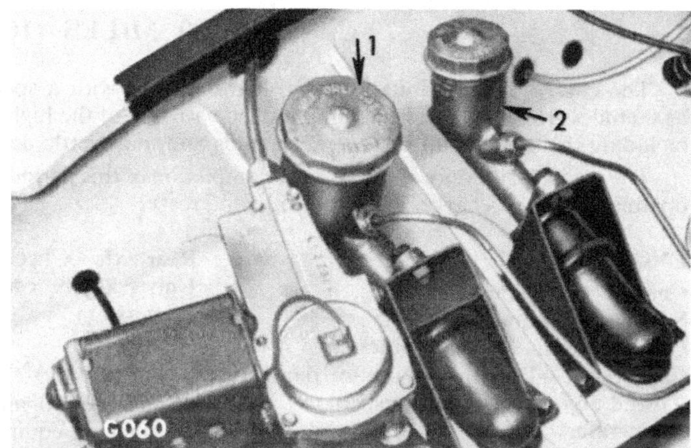

Fig. 71 (*upper*)   Fig. 72 (*lower*)

1,000 MILES (1600 KM.) (Free Service)

The engine sump is initially filled at the factory with a special running-in oil which should be drained after completing the first 1,000 miles (1600 km.) and the sump refilled with one of the high grade oils recommended. During this period many of the components including the brakes, fan belt, gaskets, studs and nuts, settle down, thus necessitating slight adjustment and an overall check.

The owner is, therefore, urged at the completion of this period to return the vehicle to the selling dealer who will perform the following operations free-of-charge, except for oil and grease.

ENGINE

Coolant—Check level
Sump—Drain and refill
Cylinder head—Check tightness
Carburettor—Top up carburettor dampers and adjust engine idling speed
Accelerator control linkage and pedal fulcrum—Oil
Fan belt—Adjust tension
Valves—Adjust clearances
Mounting bolts—Check tightness
Manifolds—Check tightness
Oil filter—Check for oil leaks

CLUTCH AND CONTROLS

Master cylinder—Top up
Hydraulic pipes—Check for leakage

TRANSMISSION

Gearbox, Overdrive—Check level and top up

Rear axle—Check level and top up
Universal joint coupling bolts—Check tightness

STEERING AND SUSPENSION

Front wheel alignment—Check with aid of tracking equipment
Rear wheel alignment—Check by condition of tyre tread
Steering unit attachments and "U" bolts—Check for tightness
Tie rods and levers—Check for tightness
Lower steering swivels—Oil

BRAKES AND CONTROLS

Handbrake cable and linkage—Lubricate
Hydraulic pipes—Check for leaks, chafing and for hose clearance
Master cylinder—Check level and top up
Brake shoes and handbrake cable—Adjust as necessary

ELECTRICAL EQUIPMENT

Battery—Check and adjust level. Check charging rate
Generator and starter motor—Check fixing bolts for tightness
Distributor—Lubricate and adjust points
Headlamp—Check alignment and adjust if required
Lights, heater, screen washer, wipers and warning equipment—Check operation

WHEELS AND TYRES

Wheel nuts—Check tightness
Tyres—Check and adjust pressures

BODY

Door strikers, locks and hinges—Oil and check operation
Body mounting bolts—Check tightness
Door handles, controls and windscreen—Wipe clean.

6,000 MILES (10,000 KM.)

In addition to the periodic checks:—

Change Engine Oil (Figs. 67 and 73)

The use of approved engine lubricants, listed on pages 70 and 71 is important when changing the engine oil every 6,000 miles (10,000 km.), or six months, whichever is the earlier.

Reduce the period between changes according to the severity of the following unfavourable conditions:—

1. Dusty roads.

2. Short journeys involving frequent stop/start driving, particularly during cold weather when greater use is made of the choke control.

If the vehicle is used for competition or sustained high speed work the use of higher viscosity oil is recommended because of the increased oil temperature.

In all cases where unfavourable operating conditions exist, you are advised to consult your Standard-Triumph Distributor or Dealer.

Procedure

Remove the plug (2), Fig. 73, to drain the oil.   Replace the plug, refill to the correct level with an approved oil (Pages 70 and 71) via the cap, Fig. 67.

Top-up Gearbox/Overdrive Unit (Fig. 73)

With the vehicle standing on level ground, remove the oil level plug (1) and using a suitable dispenser such as a pump type oil can with flexible nozzle, filled with an extreme pressure (Hypoid) lubricant, top up the gearbox and overdrive until the oil is level with the bottom of the filler plug threads.

Allow surplus oil to drain away before refitting the level plug and wiping clean.   A transfer hole permits the units to attain a common level.

Fig. 73

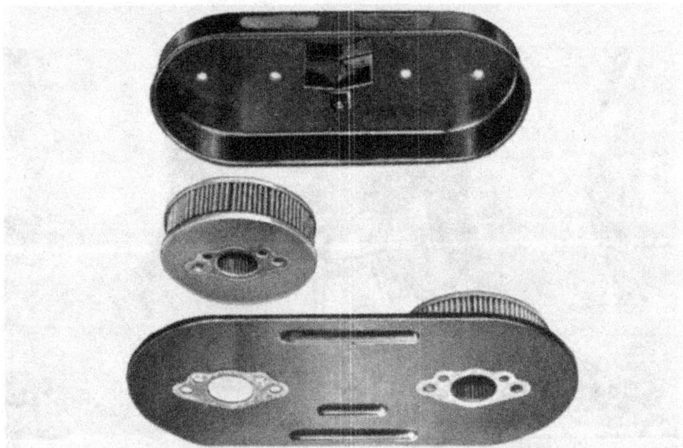

Fig. 74 (*upper*)　　　Fig. 75 (*lower*)

Top-up Rear Axle (Fig. 74)

Remove the oil level plug (arrowed) and, using the same dispenser as used for topping-up the gearbox, and the same oil, *i.e.*, extreme pressure (Hypoid) lubricant, top up the rear axle until the oil is level with the bottom of the filler plug threads.

Allow surplus oil to drain away before refitting the level plug and wiping clean.

Air Cleaners (Fig. 75)

Remove and de-dust the paper elements.

Remove the bolts attaching the assembly to the carburettor flanges, and detach the assembly.

Remove the centre bolt from the body, take off the cover plate and lift out the element. Clean out the body and use a foot pump, low pressure air line, or soft brush to remove foreign matter from between the folds of the paper element.

Reverse the foregoing procedure to assemble the unit, renewing the gaskets as necessary.

If the engine is operating in dusty conditions, clean the filters more frequently.

Valve Rocker Clearances (Fig. 95)

Check the valve rocker clearances and, if required, reset them to 0·010″ (0·25 mm.) (cold). See page 61.

Carburettor Dampers (Fig. 76)

Unscrew the hexagon plug from the top of each carburettor and withdraw the plug and damper assembly. Top up the damper chambers with the current grade of engine oil. The oil level is correct when, utilizing the damper as a dipstick, its threaded plug is ¼″ (6 mm.) above the dashpots when resistance is felt. Refit the damper and hexagonal plug. Using an oil can, apply oil to the throttle and choke control linkages.

Fan Belt Adjustment (Fig. 77)

The fan belt should be sufficiently tight to drive the water pump and generator without unduly loading the bearings.

Adjust the belt by slackening the adjusting bolt (1), and the generator pivot bolts (2). Pivot the generator until the belt can be moved ¾″ to 1″ (19/25 mm.) sideways at the mid-point of its longest run. Maintaining the generator in this position, securely tighten the adjusting bolt and the two pivot bolts.

Sparking Plugs

Remove the sparking plugs for cleaning and re-set the gaps to 0·025″ (0·63 mm.). Clean the ceramic insulators and examine them for cracks or other damage likely to cause "HT" tracking. Test the plugs before refitting and renew those which are suspect.

Fig. 76 (*upper*)   Fig. 77 (*lower*)

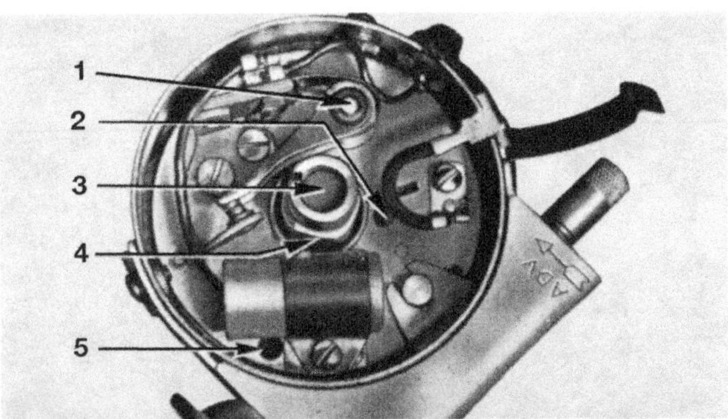

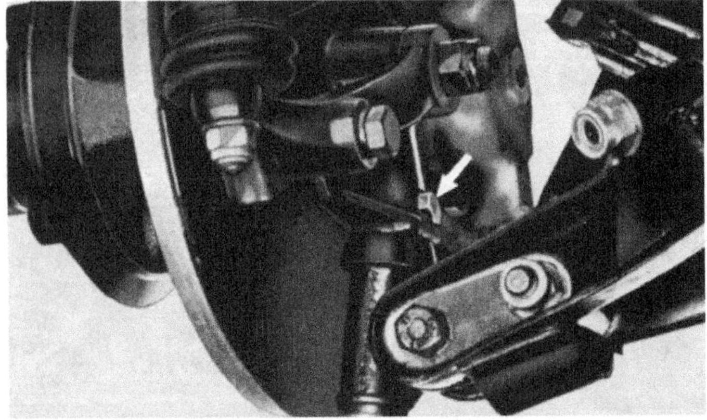

Fig. 78 (*upper*)   Fig. 79 (*lower*)

Distributor (Fig. 78)

Release the clips and remove the distributor cap and rotor arm. Apply a few drops of thin oil to points (1), (2) and (3). Lightly grease the cam surface (4) and inject approximately 5 c.c (one teaspoonful) of engine oil through the hole (5).

Clean and adjust the contact breaker points as described on page 60.

Engine Oil Filler Cap (Fig. 67)

Clean the filler cap with petrol or paraffin, allow to dry and replace. **Do not oil.**

Engine Slow Running

Check and adjust if necessary.   Page 63.

Steering Lower Swivels (Fig. 79)

Raise the front road wheels clear of the ground, remove the plug (shown arrowed) and fit a screwed grease nipple. Apply a grease gun filled with HYPOID OIL and pump the gun until oil exudes from the swivel. Remove the nipple, refit the plug Repeat, with the opposite steering swivel.

Brakes

Examine brake pads and shoes.   Adjust or, if necessary, renew.   Page 67.

Clutch and Brake Hoses

Examine for leakage and renew defective hoses.   Ensure they have adequate clearance to prevent chafing against other components.

Handbrake Cable Guides

Apply grease around the cable guides (Fig. 80) and the compensator sector (Fig. 81).

Tyres and Wheel Nuts

Examine tyres (page 24); take appropriate remedial action if necessary and have the front and rear wheel alignment checked at your Standard-Triumph Distributor or Dealer.   Check the wheel nuts for tightness.

Door Strikers, Locks and Hinges

Oil and check operation.

Door Handles, Controls and Windscreen

Wipe clean.

Electrical

Check the operation of all electrical equipment.

Fig. 80 (*upper*)   Fig. 81 (*lower*)

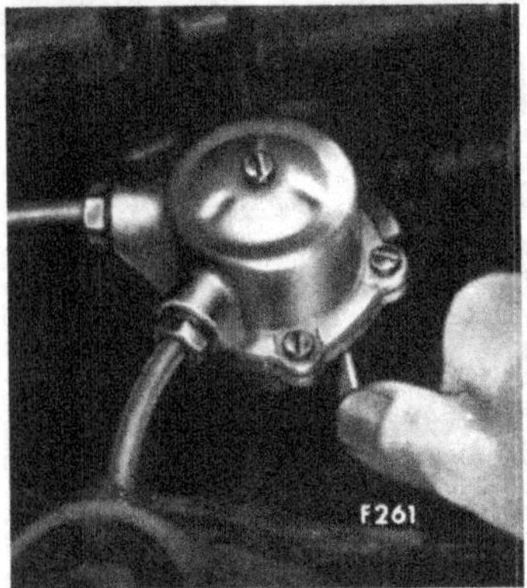

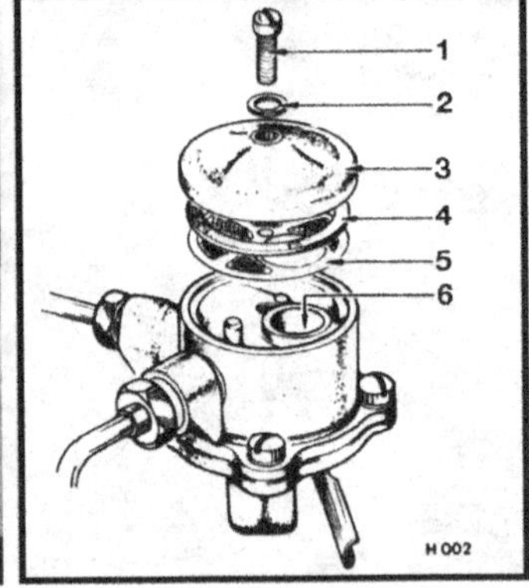

Fig. 82 (*left*)   **Fig. 83** (*upper*)   **Fig. 84** (*right*)

12,000 MILES (20,000 KM.)

Every 12,000 miles (20,000 km.) or every 12 months, whichever is earlier, carry out the work listed under 6,000 miles (10,000 km.) and the following additional work:

Oil Filter (Fig. 83)

Unscrew the old filter from the cylinder block and replace it with a new one.   Ensure an oil-tight seal by cleaning and smearing the joint faces with oil before screwing the new filter unit tightly home.

Fuel Pump (Figs. 82 and 84)

Clean out the fuel pump.

Access to the fuel pump bowl and filter is gained by unscrewing the bolt (1) and removing the domed cover (3). Lift the filter gauze (5) from its seating and wash it in petrol.

Using a small screwdriver, loosen the sediment in the bowl and remove it with compressed air.   A foot pump used for tyre inflation is ideal for this purpose.

IMPORTANT.   Do not disturb or damage the non-return valve (6) during this process.

Renew the cork gasket (4) if this has hardened or is broken. Assemble the filter gauze (5) into its seating, taking care to place the gauze face downwards so that it can be removed easily when required.

Sparking Plugs

Remove and discard the sparking plugs. When fitting replacements, ensure that they are of the correct type and that the gaps are set to 0·025″ (0·63 mm.). Re-connect the leads in the order shown in Fig. 99.

Crankcase Breather Valve (Figs. 85 and 86)

Disengage the clip (6) from the valve body (1) and lift out the diaphragm (4) and spring (2). Clean the components by swilling them in methylated spirits (denatured alcohol). Ensure that the breather pipe is clean and serviceable. Reverse the dismantling sequence to re-assemble.

NOTE. When the breather valve is cleaned, remove the oil filler cap and check that the breather hole is unobstructed and that the joint washer is serviceable.

Generator Rear Bearing (Fig. 87)

Inject a few drops of medium viscosity oil through the hole in the rear bearing housing.

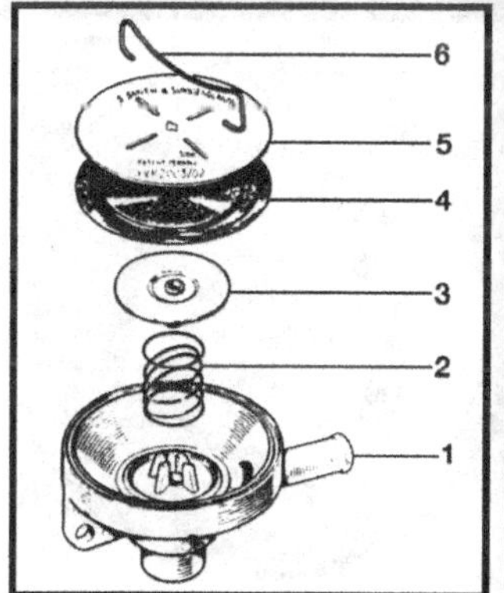

Fig. 85 (*left*)   **Fig. 86** (*upper*)   **Fig. 87** (*right*)

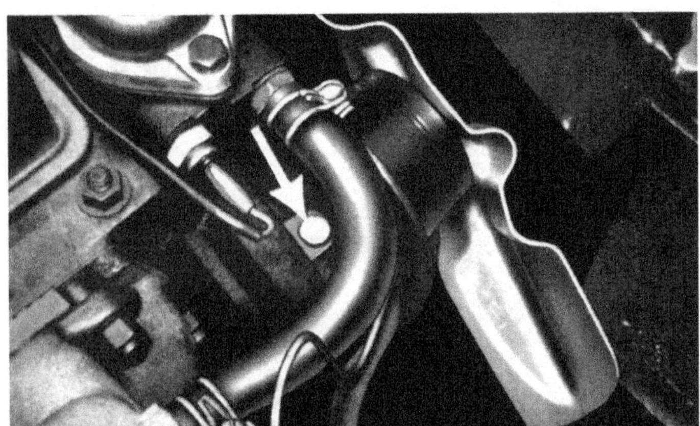

Fig. 88 (*left*)   **Fig. 89** (*upper*)   **Fig. 90** (*right*)

Water Pump (Fig. 89)

Remove the plug and fit a screwed grease nipple.   Apply a grease gun, giving five strokes only.   Remove the nipple and refit the plug.

Steering Unit (Fig. 88)

Remove the plug from the top of the unit and fit a screwed grease nipple.   Apply the grease gun and give five strokes only. Remove the nipple and refit the plug.   Over-greasing can cause damage to the rubber bellows.

Rear Brake Shoes (Fig. 90)

Jack up the rear of the car and remove both road wheels and brake drums. Examine the brake linings and if one or more shoes are excessively worn or contaminated with oil or grease, renew the complete set to ensure braking stability.   Using a high pressure air line, or a foot pump, blow all loose dust from the mechanism and, using a clean dry cloth, wipe the dust from the inside of the drums.   Avoid touching the braking surfaces with greasy hands.

Refit the brake drums and road wheels. Re-adjust the brakes, and remove the jack.

Front Brake Pads (Fig. 91)

Jack up the front of the car, remove the front wheels and examine the brake pads for wear.   Renewal is necessary when the lining material bonded to the steel backing plate is reduced to a thickness of $\frac{1}{8}''$ (3·0 mm.).   (See page 67).

Rear Hubs (Fig. 93)

Remove the plug and fit a screwed grease nipple. Apply a grease gun until grease exudes from the bearing. Remove the nipple and refit the plug. Repeat, with the opposite rear hub.

Front Hub Adjustment (Fig. 92)

If hub end float is apparent, jack up the front of the vehicle, remove the nave plate and road wheel, and prise off the dust cap. Remove the split pin, and tighten the nut whilst rotating the hub, until slight resistance is felt. Slacken the nut by one half flat and insert a new split pin. Replace the dust cap, road wheel and nave plate.

If the car is being used for competition work, re-pack the front hubs with grease every 12,000 miles (20,000 km.). For normal use, re-pack the front hubs with grease every 24,000 miles (40,000 km.)

Exhaust System

To prevent injury to health, check the exhaust system for leakage, damage and deterioration and rectify as necessary.

Attachments

Examine and, if necessary, tighten the following: Front and rear suspension attachments, steering connections, water pump, starter motor generator and generator pulley, oil filter and all universal couplings and bolts.

Fig. 91 *(left)*   Fig. 92 *(upper)*   Fig. 93 *(right)*

Fig. 94 (*upper*)   **Fig. 95** (*lower*)

RUNNING ADJUSTMENTS

Valve Seat Attention

Occasionally have the compression pressures checked by your Triumph Dealer. Providing the engine is functioning satisfactorily, and the compression pressures of all the cylinders are equal, you are advised not to disturb the engine.

The need for decarbonising arises when the build-up of carbon, a product of combustion, becomes excessive. If premium grade fuels and high quality lubricants are used in modern high compression engines, carbon deposit is so minimised that frequent decarbonising is unnecessary. Carbon removal may, therefore, be restricted to occasions when the cylinder head is removed for attention to the valves and seats.

Contact Breaker Points (Fig. 94)

With the cap and rotor arm removed, turn the engine until the contact breaker lever is operating on the highest point of the cam lobe, *i.e.*, gap at its widest. Slacken the fixed contact screw (2) and turn the eccentric screw (3) to obtain 0·015″ (0·4 mm.)

gap using a feeler gauge between the contacts (1), and retighten screw (2).

Renew worn or damaged points.

Valve Clearances (Fig. 95)

Adjust the rocker clearances to 0·010″ (0·25 mm.) (cold). A gauge is provided in the tool kit. Remove the plugs and rotating the engine in a running direction, proceed as follows:

Adjust Nos. 1 and 3 valves with Nos. 8 and 6 valves open.

,,	,,	5	,,	2	,,	,,	,,	4	,,	7	,,	,,
,,	,,	8	,,	6	,,	,,	,,	1	,,	3	,,	,,
,,	,,	4	,,	7	,,	,,	,,	5	,,	2	,,	,,

Cylinder Head Nuts ((Fig. 96)

Tighten the cylinder head nuts in the order shown. Slacken them by reversing the sequence.

Top Dead Centre (Fig. 97)

To obtain Top Dead Centre, turn the crankshaft until the small hole in the driving pulley coincides with the pointer attached to the timing cover and the rotor arm segment points in the direction shown on Fig. 98.

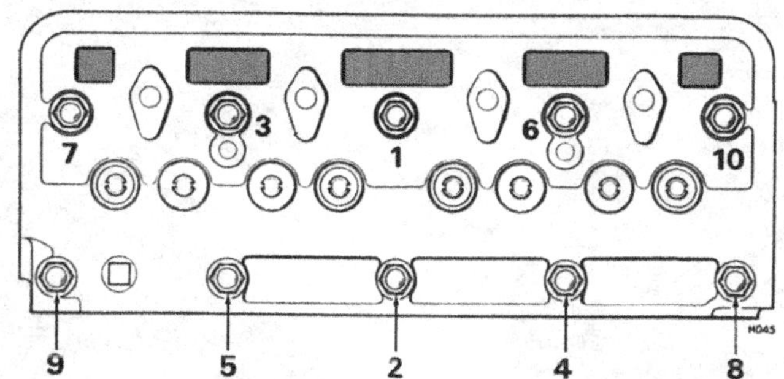

Fig. 96 *(upper)*   Fig. 97 *(lower)*

Fig. 98 (*upper*)   **Fig. 99** (*lower*)

Ignition Timing (Static)

Remove the distributor cap, adjust the contact breaker gap to 0·015″ (0·4 mm.) and set the micro-adjuster to its mid-way position.   Turn the engine until No. 1 piston is at T.D.C. on the compression stroke (see "Top Dead Centre" page 61) and the rotor arm points as shown on Fig. 98.

Mark the periphery of the pulley with a spot of white paint, 6·0 mm. to the left of the timing hole (viewed from the rear). Turn the crankshaft in a running direction almost two complete revolutions, until the white paint spot coincides with the tip of the pointer. Slacken the distributor clamp plate and switch on the ignition. Turn the distributor body slightly in either direction until the contact breaker points are seen or heard to break.   Tighten the distributor clamp plate, switch off the ignition and refit the distributor cap.

Static ignition timing (97 octane fuel) .. .. 6° B.T.D.C.

Plug Lead Positions (Fig. 99)

Ensure that the plug leads are attached to the sparking plugs as shown.   Firing order is 1, 3, 4, 2, taken in anti-clockwise order.

CARBURETTORS

Tuning Carburettors (Figs. 100 to 104)

Twin carburettor installations cannot be successfully tuned unless the general conditions of the engine, ignition and the fuel systems is satisfactory. Before attempting to tune carburettors, ensure that the engine oil filler cap and the dipstick are both secure, as air drawn into the crankcase will affect the engine slow running.

Remove the air cleaners and run the engine until it reaches normal operating temperature. Slacken the clamping bolts on the throttle spindle connections (2). Close the throttles fully by unscrewing the idling adjustment screws (1) and then open by screwing down the screws one and a half turns.

Remove the suction chambers and pistons. Screw up the jet adjusting nuts (3) until each jet is flush with the bridge of its carburettor, or as near to this as possible. Replace the pistons and suction chamber assemblies and check that the pistons fall freely. Turn down the jet adjusting nuts two complete turns (12 flats).

Restart the engine and adjust the throttle adjusting screws (1) to give the desired idling speed by moving each throttle adjusting screw an equal amount. Using a length of 0·3″ (8 mm.) approx. bore tubing, listen to the hiss in the intakes and adjust the throttle adjusting screws until the intensity of the hiss is similar at both intakes. This will synchronize the throttles.

Set the pins of the throttle interconnecting levers (2) 0·015″ (0·4 mm.) from the lower edge of the forks (Fig. 100), ensuring that there is $\frac{1}{32}$″ (0·8 mm.) side play between the interconnecting levers and the throttle nuts.

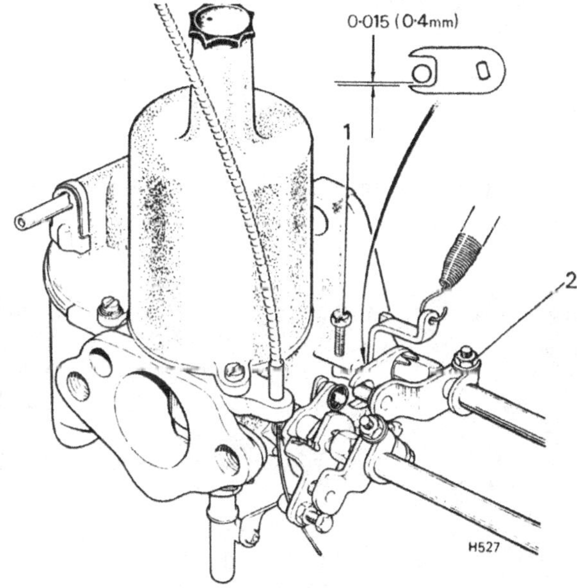

Fig. 100

When this is satisfactory, adjust the mixture by screwing both jet adjusting nuts (3) up or down by the same amount to give the fastest idling speed consistent with even firing. Press the jets (4) upwards during adjustment to ensure that they are in contact with the adjusting nuts.

Should the engine speed increase as the jets are adjusted, unscrew the throttle adjusting screws a little, each by the same amount, to reduce the speed.

Lift the piston of the front carburettor approximately $\frac{1}{32}''$ (·75 mm.) :

(a) If the engine speed increases, the mixture strength of the front carburettor is too rich ;

(b) If the engine speed immediately decreases, the mixture strength of the front carburettor is too weak ;

(c) If the engine speed momentarily increases very slightly, the mixture strength of the front carburettor is correct.

Repeat the operation at the rear carburettor and, after adjustment, re-check the front carburettor, since the two carburettors are interdependent.

Fig. 101

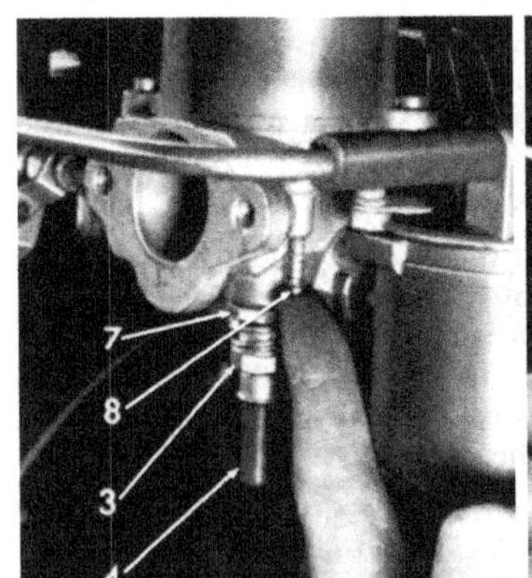

Fig. 102

Fig. 103

When the mixture is correct the exhaust note should be regular and even. If it is irregular with a splashy type of misfire and with a colourless exhaust, the mixture is too weak. If there is a rhythmical type of misfire in the exhaust beat together with a blackish exhaust the mixture is too rich.

Jet and Throttle Interconnection Adjustment (Figs. 104 and 105)

With the choke control fully "IN", the engine warm and idling on a closed throttle, adjust the screw (5) to give a clearance of 0·015″ (0·4 mm.) between the end of the screw and rocker lever.

Always check this adjustment when the throttle stop screw (1) is altered.

Cleaning Suction Chamber and Piston

At approximate intervals of twelve months, detach the piston unit. Clean the piston and the inside bore of the suction chamber. Re-assemble dry except for a few spots of thin oil on the piston rod.

Replenish the damper reservoir.

Cleaning Float Chambers

Periodically remove the fuel feed pipe, float chamber lid, and float assembly, from each carburettor. Remove all sediment from the float chambers and re-assemble the carburettors.

Fig. 104

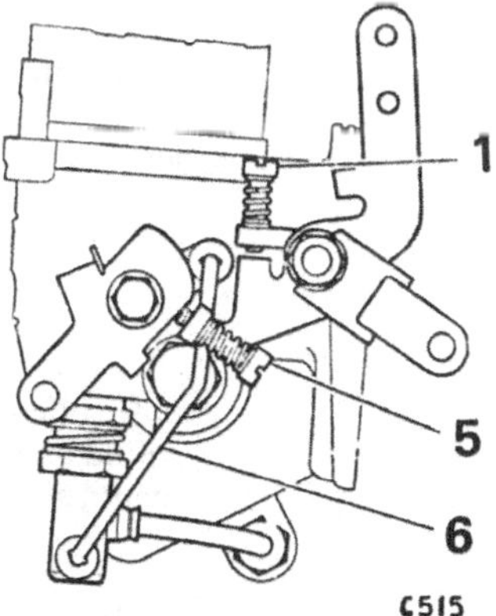

Fig. 105

Fig. 106

Float Chamber Fuel Level (Fig. 107)

The level of fuel in the float chamber is adjusted by setting the float lever on the float chamber lid, as follows:—

Disconnect the fuel feed pipe and remove the float chamber lid.

Invert the lid and, with the float lever resting on the needle valve, measure the gap between the lever and lower lid face as shown.

If necessary, bend the float lever to obtain the correct setting.

Refit the float chamber lid, and re-connect the fuel pipe.

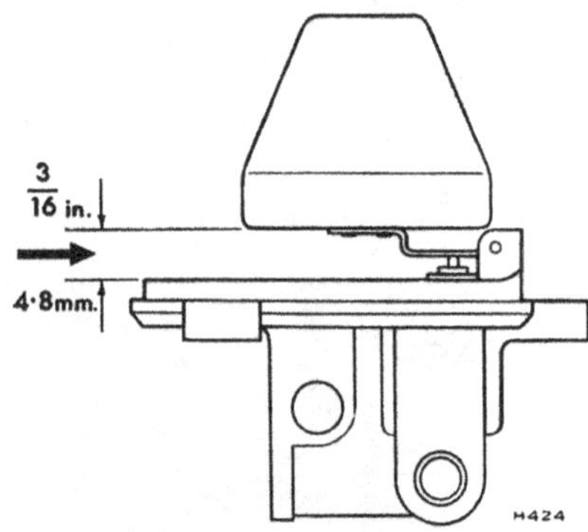

Fig. 107

Jet Centralising (Figs. 105 and 106)

If the suction piston is lifted and released, it should fall freely and hit the inside "jet bridge" with a soft metallic click when the jet adjusting nut (3) is screwed to its topmost position.

If a click is audible only when the jet is in the fully lowered position, the jet should be centralised as follows:—

Disconnect the rod between the jet lever and jet head (6, Fig. 105).

Holding the jet (4) in its upper position, slacken the gland nut (7, Fig. 102) and move the jet assembly laterally until the jet is concentric with the needle, then tighten the gland nut. The piston should now fall freely and hit the jet bridge with a soft metallic click.

Lower the jet and again lift and release the piston, noting any difference in the sound of impact. If a sharper impact sound results, repeat the centralising operation to achieve identical sounds with the jet raised and lowered.

Re-connect the jet lever (6), replenish the dampers and tune the carburettors before replacing the air cleaners.

BRAKES

Hydraulically operated disc brakes are fitted to the front wheels and drum brakes to the rear wheels. Pressure applied to the brake pedal is transferred to the hydraulic system, to operate the brakes. The handbrake is connected by cables and compensating mechanism to levers incorporated in the rear brake backing plates. Applying the handbrake operates the rear brakes only, independently of the hydraulic system.

Front Brake Adjustment

The disc brakes fitted to the front of the vehicle are self-adjusting. Replacement shoe pads should be fitted when the linings are reduced to approximately $\frac{1}{8}''$ (3 mm.) thickness.

NOTE. Under no circumstances allow the pads to wear below $\frac{1}{16}''$ (1·5 mm.) thickness. They should therefore be renewed if of insufficient thickness to ensure safe braking for a further 6,000 miles (10,000 km.).

Periodically check the tightness of the caliper securing bolts (Fig. 108).

Pad Renewal (Fig. 109)

Pad renewal may be effected without having to bleed the brake system.

1. Jack up the front of the car and remove the road wheels.
2. Withdraw the retaining pins (1) and remove the pad retainers (2).
3. Remove the pads (3) from the calipers.
4. Push the pistons to the bottom of their cylinders and fit new pads.
5. Refit the pad retainers and secure with the retaining pins.
6. Pump the foot pedal until solid resistance is felt.

Rear Brakes (Fig. 110)

Brake shoes, contaminated with oil or grease, are detrimental to brake efficiency. Should a brake be so affected, the drum and backing plate must be thoroughly cleaned with petrol, and the brake shoes renewed. When renewing or replacing shoes, the pull-off springs, behind the shoes, must hook through the correct holes, as shown.

Fig. 108   Fig. 109

Adjustment (Fig. 111)

Excessive foot and handbrake travel indicates the need for rear brake adjustment.

Each rear brake is provided with an adjuster on the rear of the backing plate. To adjust the shoes, turn the adjuster (2, Fig. 113) clockwise until the shoes are hard against the drum: then slacken the adjuster by one notch increments until the drum is free to rotate.

NOTE. There is a constant drag on the rear wheels caused by the action of the differential and axle oil. Do not confuse this with brake drag.

Handbrake Adjustment (Fig. 111)

Adjustment of the brake shoes re-adjust the handbrake mechanism. If cable slackness remains, re-adjust the handbrake clevis on each rear brake assembly. Do not overtighten the cable.

1. Unhook the spring (1).
2. Remove split pin and clevis pin (2).
3. Slacken locknut (3) and adjust clevis nut (4).
4. Adjust to suit and re-assemble.

Procedure for Bleeding (Figs. 112 and 113)

Should a pipe joint have been uncoupled or the brake pedal become spongy, bleed the system from each wheel cylinder in turn starting with the front left-hand side wheel and proceeding to each wheel in turn, from the farthest wheel to that nearest to the master cylinder, *i.e.*, in anti-clockwise sequence (right-hand steering). Start at the front right-hand side and proceed in a clockwise sequence (left-hand steering). (The clutch is bled from the operating cylinder on the left-hand side of the gearbox).

Fig. 110

Fig. 111

With the aid of a second operator, proceed as follows:

1. Tighten the rear brake shoes hard against the drum, to reduce the space in the cylinders.

2. Ensure that the reservoir is topped up to the level marked on the master cylinder.

3. Wipe clean the bleed nipple and attach to it a short length of small bore tubing. Allow the tube to hang in a clean container partially filled with hydraulic fluid, so that its end is below the level of the fluid.

4. Unscrew the bleed nipple one half turn.

NOTE. During bleeding, the reservoir fluid level falls rapidly. Ensure that the level does not fall below half full, by constantly replenishing with fluid that has been stored in a container sealed from atmosphere. Immediately bleeding is completed, re-seal residual fluid in the container as exposure lowers the boiling point.

5. Operate the pedal with a succession of rapid long and short strokes as follows: Push the pedal through its full stroke, followed by two or three short rapid strokes; then allow the pedal to return to its stop unaided (foot removed). Observe the fluid being discharged into the glass container and when all bubbles have ceased to appear, bleeding is complete. Securely tighten the bleed screw and remove the tubing from the nipple.

6. Top up the master cylinder with hydraulic fluid, adjust the brake shoes, and road test vehicle.

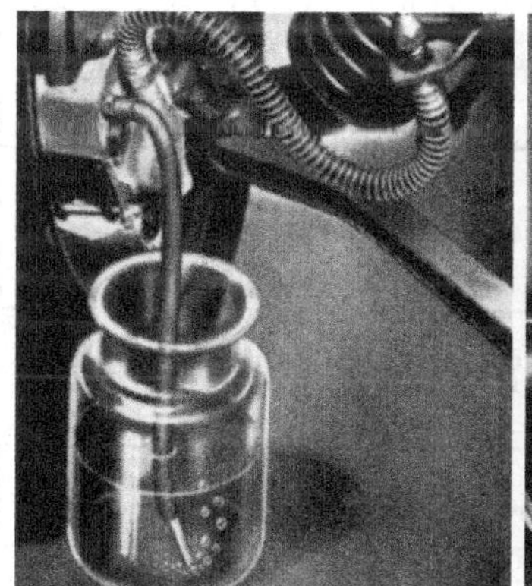

Fig. 112

Fig. 113

RECOMMENDED LUBRICANTS (BRITISH ISLES) - ANTI-FREEZE SOLUTIONS

(The products recommended are not listed in order of preference)

COMPONENTS		BP	CASTROL	DUCKHAM'S	ESSO	MOBIL	PETROFINA	REGENT	SHELL
ENGINE CARBURETTOR DAMPERS AND OIL CAN		Super Visco-Static 20W-50	Castrol GTX	Duckham's Q20-50	Esso Extra Motor Oil 20W/50	Mobiloil Super SAE 10W/40 or Mobiloil Special 20W/50	Fina Multigrade Motor Oil SAE 20W/50	Havoline Motor Oil 20W/50	Shell Super Motor Oil 100
LOWER SWIVELS GEARBOX REAR AXLE OVERDRIVE		Gear Oil SAE 90 EP	Castrol Hypoy	Duckham's Hypoid 90	Esso Gear Oil GP 90/140	Mobilube GX 90	Fina Pontonic MP SAE 90	Multigear Lubricant EP90	Shell Spirax 90EP
FRONT HUBS BRAKE CABLES GREASE GUN		Energrease L2	Castrolease LM	Duckham's LB 10	Esso Multi-Purpose Grease H	Mobilgrease MP	Fina Marson HTL2	Marfak All-Purpose	Shell Retinax A
CLUTCH AND BRAKE RESERVOIRS		CASTROL GIRLING BRAKE AND CLUTCH FLUID CRIMSON. WHERE THIS PROPRIETARY BRAND IS NOT AVAILABLE, OTHER FLUIDS WHICH MEET SAE 70R3 SPECIFICATION MAY BE USED							
APPROVED ANTI-FREEZE SOLUTIONS	Smith's Bluecol	BP Anti-Frost	Castrol Anti-Freeze	Duckham's Anti-Freeze	Esso Anti-Freeze	Mobil Permazone	Fina Thermidor	Regent PT Anti-Freeze	Shell Anti-Freeze
	WHERE THESE PROPRIETARY SOLUTIONS ARE NOT AVAILABLE, OTHERS WHICH MEET BSI 3151 or 3152 SPECIFICATION MAY BE USED								

(The products recommended are not listed in order of preference)

COMPON-ENT	AIR TEMP. °C.	AIR TEMP. °F.	API DESIG-NATION	BP	CASTROL	DUCKHAM'S	ESSO	MOBIL	PETROFINA	SHELL	TEXACO CALTEX
ENGINE CARBUR-ETTOR DAMPERS AND OIL CAN	over 30	over 80	MM or MS	Visco-Static · Energol SAE 30 · Visco-Static Long-Life · Castrolite 10W/30	Castrol 30 HD · Castrol XL 20W/40	Q20-50	Esso Extra Motor Oil 20W/40	Mobiloil Special 10W/30 or Mobiloil Super SAE 10W/40	Fina Multigrade Motor Oil SAE 20W/50	Shell Super Motor Oil	Havoline 30 · Havoline 10W/30 · Havoline 20W/40
	0 to 30	30 to 80	MM or MS	Energol SAE 20W	Castrol 20 HD		Esso Extra Motor Oil 10W/30				Havoline 20/20W
	below 0	below 30	MM or MS	Energol SAE 10 W	10 HD	Q5500			Fina Multigrade Motor Oil SAE 10W/30		Havoline 10W
LOWER SWIVELS GEARBOX REAR AXLE OVER-DRIVE	over 0	over 30	GL 4	Gear Oil SAE 90 EP	Castrol Hypoy	Duckham's Hypoid 90	Esso Gear Oil GP 90	Mobilube GX 90	Fina Pontonic MP SAE 90	Shell Spirax 90 EP	Multigear Lubricant EP 90
	below 0	below 30	GL 4	Gear Oil SAE 80 EP	Castrol Hypoy Light	Duckham's Hypoid 80	Esso Gear Oil GP 80	Mobilube GX 80	Fina Pontonic MP SAE 80	Shell Spirax 80 EP	Multigear Lubricant EP 80
FRONT HUBS BRAKE CABLES GREASE GUN				Energrease L2	Castrolease LM	Duckham's LB 10	Esso Multi-Purpose Grease H	Mobilgrease MP	Fina Marson HTL 2	Shell Retinax A	Marfak All Purpose
CLUTCH AND BRAKE RESERVOIRS					CASTROL GIRLING BRAKE AND CLUTCH FLUID CRIMSON. WHERE THIS PROPRIETARY BRAND IS NOT AVAILABLE, OTHER FLUIDS WHICH MEET SAE 70R3 SPECIFICATION MAY BE USED						
APPROVED ANTI-FREEZE SOLUTIONS	Smith's Bluecol		BP Anti-Frost	Castrol Anti-Freeze	Duckham's Anti-Freeze	Esso Anti-Freeze	Mobil Permazone	Fina Thermidor	Shell Anti-Freeze	Startex	
			WHERE THESE PROPRIETARY BRANDS ARE NOT AVAILABLE, OTHERS WHICH MEET BSI 3151 or 3152 SPECIFICATION MAY BE USED								

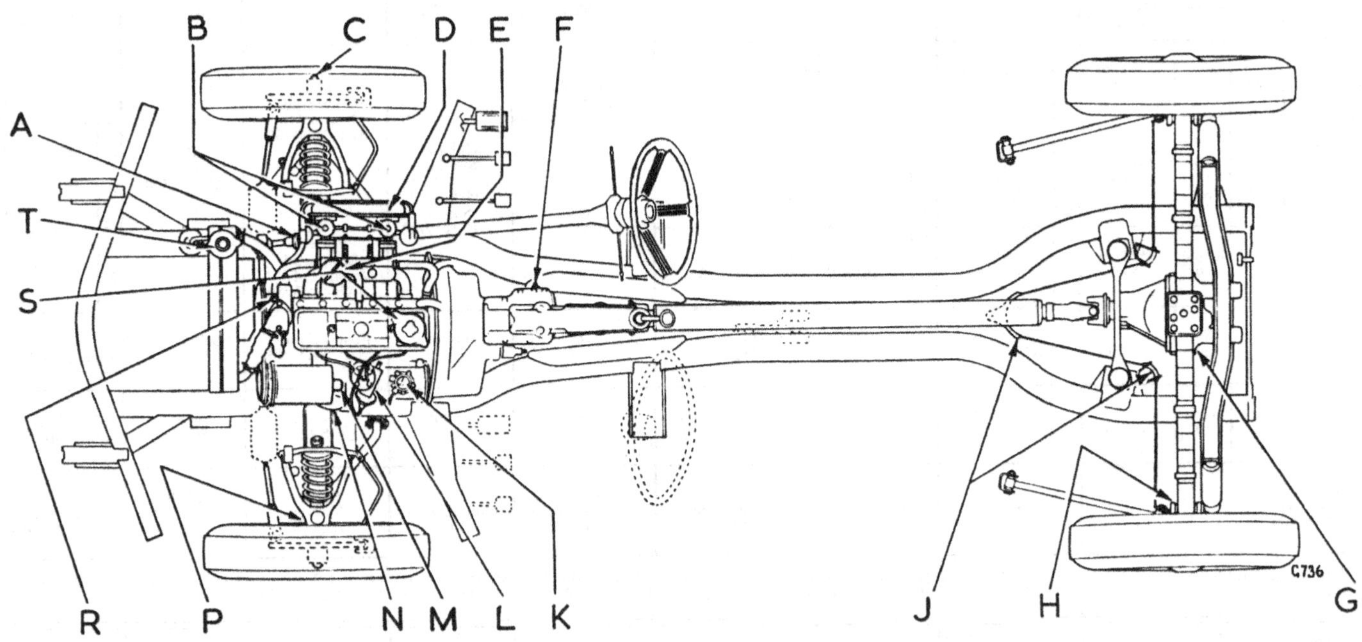

Fig. 114

Chart Ref.	Items	Details	Page Ref.		Intervals × 1000			
					Miles	Kms.	Miles	Kms.
A	Steering unit	Grease — 5 strokes	58	—			12	20
B	Carburettor dampers	Top up	53	—	6	10		
C	Front wheel hubs	Check and adjust	59	—			12	20
D	Air cleaners	Clean	52	—	6	10		
E	Crankcase breather valve	Clean	57	—			12	20
F	Gearbox	Top up	51	—	6	10		
G	Rear axle	Top up	52	—	6	10		
H	Rear hubs	Grease	59	—			12	20
J	Handbrake cable guides	Grease	55	—	6	10		
K	Fuel pump	Clean	56	—			12	20
L	Distributor	Oil	54	—	6	10		
M	Generator rear bearing	Oil	57	—			12	20
N	Oil cleaner	Replace	56	—			12	20
P	Lower steering swivels	Oil	54	—	6	10		
R	Water Pump	Grease	58	—			12	20
S	Engine sump	Top up oil level	47	Daily				
		Drain and refill	51	—	6	10		
		Clean filler cap	54	—	6	10		
T	Radiator	Top up	47	Weekly				
	Brake and clutch master cylinders	Top up	49	Monthly				
	Battery	Top up	48	Monthly				

GENERAL SPECIFICATION

Engine

Number of cylinders	4	
Bore of cylinders	2·9 in.	73·7 mm.
Stroke of crankshaft	2·99 in.	76·0 mm.
Cubic capacity	79·2 in.³	1296 c.c.
Piston area	26·5 in.²	171·0 cm.²
Compression ratio	9·0 : 1	

Valve rocker clearances (cold) — 0·010 in. — 0·25 mm.

Valve timing — Inlet and exhaust valve equally open at T.D.C.

Maximum power — 75 b.h.p. at 6000 r.p.m.

Maximum torque — 900 lb.in. (10·37 kg/m.) at 4000 r.p.m.

Equivalent to B.M.E.P. — 144 p.s.i. (10·1 Kg/cm²)

Lubrication (Engine)

Pump — High capacity eccentric lobe type.

Filter — Replaceable full-flow type.

Pressure at 2000 r.p.m. — 40-60 p.s.i. (2·8-4·2 Kg/cm²)

Cooling System

Circulation — "No loss" system incorporating plastic overflow reservoir. Pump, driven by "Vee" belt; thermostatically controlled flow.

Fan — Four blade 12·125 in. dia. (308 mm.).

Fuel System — Tank above rear axle. No reserve tap.

Pump — A.C. mechanically operated diaphragm type.

Carburettor — Twin H.S.2 S.U.

Needle size — B.O.

Manifolds — Separate inlet and exhaust. Water heated inlet manifold

Air cleaners — Twin replaceable paper elements.

Crankcase breathing — Closed circuit—controlled by emission valve between rocker cover and manifold.

Ignition System

Coil — Lucas HA12.

Distributor — Delco-Remy.

Contact breaker gap — 0·015 in. (0·4 mm.).

Rotation — Anti-clockwise.

Firing order — 1-3-4-2

Sparking plugs — Champion N.9.Y

Gap — 0·025 in. (0·64 mm.).

Ignition timing (static) — 6° B.T.D.C.

Electrical System

Voltage — 12

Polarity — Negative earth

Fuses	35 amp.
Battery—Type	Lucas D.9
Capacity at 20 hr. rate	40 amp. hour
Plates per cell	9
Normal charging rate	3·5 amps.
Generator	Lucas C.40-1.
Output	264 watts.
Control box	Lucas RB.340.
Flasher unit	Lucas FL.5.
Fuel and temperature indication	Bi-metal type, 10 volt system.
Oil pressure switch operating pressure	4·5/7·5 p.s.i. (0·32/0·53 kg/cm^2).

Transmission

Clutch	Diaphragm type. 6½ in. dia. (165 mm.)
Gearbox	4 speed constant mesh. Synchromesh on 2nd, 3rd and Top.

	Top	3rd	2nd	1st & Rev.
Ratios	1	1·39	2·16	3·75
Overall ratios	4·11	5·73	8·87	15·4

Overdrive (optional)	Laycock de Normanville. Operates on 3rd and Top gears. Ratio 0·802:1
Rear axle	Swing axle. Centre rubber mounted to chassis. Hypoid bevel drive. Ratio 4·11 to 1.

Wheels	Pressed steel disc type 3½ D rim. Wire spoked wheels (optional).
Tyres	5·20S-13 Tubeless. 1·45-13 Radial ply (optional).
Pressures	See Page 25.

Brake System — Girling hydraulic. Centrally mounted "fly-off" handbrake coupled mechanically to rear wheels only.

Front	Caliper disc 9 in. dia. (229 mm.).
Rear	Drum 7 in. dia. (178 mm.) × 1¼ in. (31·7 mm.) wide.

Front lining area	14·8 in^2	95 cm^2
Front swept area	150 in^2	962 cm^2
Rear lining area	34 in^2	225 cm^2
Rear swept area	55 in^2	353 cm^2
Total lining area	48·8 in^2	314 cm^2
Total swept area	205 in^2	1316 cm^2
Maximum retardation	0·98 g. equivalent to stopping from 30 m.p.h. (48 k.p.h.) in 31 ft. (9·45 m.)	

Chassis Data

Frame	Double backbone of closed channel section with channel outriggers.	
Wheelbase	6 ft. 11 in.	2110 mm.
Track—Front	4 ft. 1 in.	1245 mm.
—Rear	4 ft.	1220 mm.
Ground clearance	5 in.	125 mm.

GENERAL SPECIFICATIONS

Turning circle	24 ft.	7·3 m.
Steering unit	Rack and pinion 3¾ turns lock to lock	

Suspension

Front	Low periodicity independent suspension system with rubber bushed wishbone pivots. Patented screwed bottom bush and top ball joint swivels. Coil springs controlled by telescopic type direct acting dampers and anti-roll bar. Taper roller bearings in hub.
Rear	Swing axle type independent suspension, transverse leaf spring and radius rods. Ball and needle roller bearings in hubs.

Capacities

	Imperial	Metric	U.S.A.
Fuel tank	8·25 galls.	37·6 litres	9·9 galls.
Engine sump	8 pints	4·5 litres	9·6 pints
Gearbox	1·5 pints	0·85 litres	1·8 pints
Gearbox and Overdrive	2·38 pints	1·35 litres	2·85 pints
Rear axle	1 pint	0·57 litres	1·2 pints
Cooling system (inc. water bottle)	7 pints	4·0 litres	8·4 pints
Cooling system (inc. water bottle) with heater	8 pints	4·5 litres	9·6 pints

Exterior Dimensions

Overall length	12 ft. 3 in.	373 cm.
Width	4 ft. 9 in.	145 cm.
Height with load erected (unladen)	3 ft. 11·5 in.	120·5 cm.
Height with load folded (unladen)	3 ft. 8·5 in.	112·5 cm.

Weight (approx.)

Dry (excluding extra equipment)	14 cwt.	712 Kg.
Complete (inc. fuel, oil, water and tools)	14·75 cwt.	748 Kg.
Maximum gross vehicle weight	17·75 cwt.	905 Kg.

Road Speed Data

Engine speed at a road speed of	O/D Top	Top	O/D 3rd	3rd	2nd	1st & Rev.
10 m.p.h.	565	635	715	890	1375	2385
10 k.p.h.	350	395	490	550	875	1480

Road speed at 1000 r.p.m.	O/D Top gear	Top gear
	19·6 m.p.h.	15·75 m.p.h.
	31·3 k.p.h.	25·2 k.p.h.

Road speed at 2500 ft/min. (762 m/min.) piston speed in top gear   81 m.p.h.   130 k.p.h.

	Page
Accelerator	6, 7, 10
Air cleaners	52
Air distribution controls	11
Anti-freeze	31, 70, 71
Approved lubricants	70, 71
Battery	32, 48
Bleeding the hydraulic systems	68
Blower switch	10
Bodywork	28
Bonnet	15
Brake master cylinder	49
Brakes	46, 55, 58, 67
Breather valve	57
Bulbs (replacement)	39, 40, 41
Capacities	76
Carbon removal	60
Carburettor	53, 63
Care of bodywork	28
Choke control	10
Chromium plating	28
Clutch master cylinder	49
Commission number	4
Compression pressures	60
Contact breaker points	54, 60
Control box	33
Controls	6
Cooling system	29
Crankcase breather valve	57
Cylinder head	60, 61

	Page
Data	74
Dimensions	76
Dipstick	47
Direction indicators	8, 9, 39
Disc brake pads	55, 58, 67
Distributor	54
Door locks	15
Drain taps	29
Driving from new	26
Electrical circuit	42, 44
Electrical system	32
Engine lubrication	47, 51
Exhaust system	59
Facia	6, 7
Fan belt	33, 53
Firing order	62, 74
Flasher unit	37
Free service	50
Frost precautions	31
Fuel—recommended	27
—Filler cap	16
—Gauge	8, 37
—Pump	56
Fuses	35
Gauges	8, 9, 37
Gearbox lubrication	51
Gear lever	9
General specification	74

	Page
Generator	3, 57
Glass care	28
Handbrake—adjustment	68
—cable guides	55
—lever	10
Hard top	20
Headlamp—alignment	38
—light units	38
Heater—blower switch	10
—control	10
Horn push	9
H.T. cables	35, 62
Hubs—front	59
—rear	59
Ignition—distributor	54
—switch	8
—timing	62
—warning light	11
Interior care	28
Instruments and iudicators	6
Instrument illumination	8, 40
Jacking	21
Lighting switch	8
Lights selector switch	9
Locks and keys	15
Lubricants	70, 71

	Page
Lubrication chart	72
Lubrication summary	73
Luggage compartment	16
Main beam—warning light	9, 11
Master cylinders	49
Monthly checks	48
Needles	74
Number plate illumination	40
Odometer	10
Oil changes (engine)	51
Oil filter	56
Oil pressure switch	34
Oil pressure warning lamp	11
Overdrive	12, 51
Parking lamps	39
Pedals	10
Plate illumination lamp	40
Plugs	53, 57
Plug leads	35, 62
Polarity	32
Pump—fuel	56
—water	58
Radiator	29, 47
Rear axle—lubrication	52

Made in the USA
Monee, IL
07 July 2026

56551546R00046